David Urs Josef Keller

Multiscale Modeling of the Ventricles
From Cellular Electrophysiology to Body Surface Electrocardiograms

Vol. 13
Karlsruhe Transactions on Biomedical Engineering

Editor:
Karlsruhe Institute of Technology
Institute of Biomedical Engineering

Eine Übersicht über alle bisher in dieser Schriftenreihe erschienenen Bände
finden Sie am Ende des Buchs.

Multiscale Modeling of the Ventricles

From Cellular Electrophysiology to Body Surface Electrocardiograms

by
David Urs Josef Keller

Dissertation, Karlsruher Institut für Technologie
Fakultät für Elektrotechnik und Informationstechnik, 2011

Impressum

Karlsruher Institut für Technologie (KIT)
KIT Scientific Publishing
Straße am Forum 2
D-76131 Karlsruhe
www.ksp.kit.edu

KIT – Universität des Landes Baden-Württemberg und nationales
Forschungszentrum in der Helmholtz-Gemeinschaft

KIT Scientific Publishing 2011
Print on Demand

ISSN: 1864-5933
ISBN: 978-3-86644-714-1

Multiscale Modeling of the Ventricles: from Cellular Electrophysiology to Body Surface Electrocardiograms

Zur Erlangung des akademischen Grades eines

DOKTOR-INGENIEURS

an der Fakultät für

Elektrotechnik und Informationstechnik

des Karlsruher Instituts für Technologie (KIT)

genehmigte

DISSERTATION

von

Dipl.-Ing. David Urs Josef Keller

geb. in Ettenheim

Tag der mündlichen Prüfung: 9. Juni 2011
Referent: Prof. Dr. rer. nat. Olaf Dössel
Korreferent: Assoz. Prof. Dr. Gernot Plank

Acknowledgements

This work was conducted at the Institute of Biomedical Engineering which is part of the Karlsruhe Institute of Technology (KIT).
First and foremost, I would like to thank Prof. Olaf Dössel for the opportunity to work on this project under his supervision and guidance. His inexhaustible interest in my work was greatly appreciated.
I would also like to thank Prof. Gernot Plank for his interest in my work and his referee on my thesis.

Furthermore my thanks go to my clinical collaborators Dr. med Eberhard Scholz and Dr. med Sebastian Ley. Without the clinical ECG, BSPM and MRI data, this work would not have been possible.

A lot of my colleagues contributed to this work:
Both projects related to tissue conductivities (conductivity ranking and PCA based BSPM prediction) were conducted in collaboration with Frank Weber while the effects of ventricular deformation on the T-Wave morphology were investigated in cooperation with Oussama Jarrousse and Thomas Fritz.
Moreover thanks go to Martin Krüger for the cooperation in the realm of anatomical modeling and for providing some of the pictures that were used in this thesis.
Daniel Weiss told me a lot about ventricular electrophysiology and fostered my programming skills.
A particularly big share of my gratitude goes to my direct supervisor Dr.-Ing. Gunnar Seemann who showed an enormous amount of patience when answering my countless questions and during proofreading of various manuscripts.
I would also like to thank all of my colleagues for the great atmosphere and the welcomed distractions from day-to-day problems.

Finally I want to thank all students who contributed (with their Student Research Project or Master Thesis) to this work; i.e.: Stefan Bauer, Carola Otto, Pham Tri Dung, Stephan Lurz, Andreas Bohn and Raffi Kalayciyan.

Last but not least my thanks go to my parents, grandparents, friends and in particular to my girlfriend Meli who supported me throughout the course of this thesis.

Contents

Part II Methods

Part III Results

List of Abbreviations

AB	Apico-Basal
AKAP	A-Kinase Anchoring Protein
ANS	Autonomic Nervous System
AP	Action Potential
APD	Action Potential Duration
APM	Anterior Papillary Muscle
ASR	Atrial Sensitivity Ranking
AT	Activation Time
AUR	Atrial Uncertainty Ranking
AVN	Atrioventricular Node
BSPM	Body Surface Potential Map
CaMK	Calcium/Calmodulin Dependent Kinase
CC	Correlation Coefficient
CGAL	Computational Geometry Algorithms Library
CPU	Central Processing Unit
CSQN	Calsequestrin
CT	Computed Tomography
DAE	Differential Algebraic Equation
DOR	Dispersion of Repolarization
DTMRI	Diffusion Tensor Magnetic Resonance Imaging
EAD	Early After Depolarization
ECG	Electrocardiogram
EGFP	Enhanced Green Fluorescent Protein
FLASH	Fast Low Angle SHot
FRP	Functional Refractory Period
FWHM	Full Width at Half Maximum
GG	Gabriel and Gabriel
HB	His Bundle
HETERO	Fully Heterogeneous Model
HOM	Homogeneous Model
HSV	Hue Saturation and Value
ICD	Implantable Cardioverter Defibrillator

ICP	Iterative Closest Point
ISO	Isoproterenol
IV	Inter-Ventricular
JSR	Junctional Sarcoplasmic Reticulum
KO	Knockout
L-System	Lindenmayer System
LBB	Left Bundle Branch
LCCa	Alpha Subunit of the Longitudinal Calcium Channel
LCCb	Beta Subunit of the Longitudinal Calcium Channel
LF	Left Flank
LQT	Long-QT
LQTS	Long-QT Syndrome
LV	Left Ventricle
LVW	Left Ventricular Wall
MR	Magnetic Resonance
MRI	Magnetic Resonance Imaging
NE	Norepinephrine
NSR	Network Sarcoplasmic Reticulum
PC	Principal Component
PCA	Principal Component Analysis
PETSc	Portable, Extensible Toolkit for Scientific Computation
PF	Purkinje Fiber
PKA	Protein Kinase A
PLB	Phospholamban
PLM	Phospholemman
PMJ	Purkinje Muscle Junction
RBB	Right Bundle Branch
RD	Reaction Diffusion
RF	Right Flank
RMSE	Root Mean Square Error
RT	Repolarization Time
RV	Right Ventricle
RVW	Right Ventricular Wall
RyR	Ryanodine Receptor
SERCA	Sarco/Endoplasmic Reticulum Ca2+-ATPase
SM	Electrophysiological Model from Seemann *et al.*
SN	Sinus Node
SNR	Signal-To-Noise-Ratio
SQT	Short QT
SR	Sarcoplasmic Reticulum
SS	Subspace
SSFP	Steady State Free Precession
TdP	Torsade de Pointes
TDR	Transmural Dispersion of Repolarization

tECG	Transmural Electrocardiogram
TM	Transmural
TnI	Troponin I
TT04	First Revision of the Electrophysiological Model from ten Tusscher *et al.*
TT06	Second Revision of the Electrophysiological Model from ten Tusscher *et al.*
USB	Universal Serial Bus
VSR	Ventricular Sensitivity Ranking
VUR	Ventricular Uncertainty Ranking
WT	Wild-Type

List of Figures

List of Tables

1

Introduction

1.1 Motivation

Since the development of the first mathematical models for cellular electrophysiology and the following dawn of in-silico modeling of the whole heart, quantitative approaches that allow the description of bioelectric effects at both cell and tissue level have become indispensable for modern cardiological research. The ability to integrate data from different in-vitro experiments and the possibility to selectively change distinct model parameters (without collateral effects) are just two important features of quantitative in-silico modeling.

The recent but steadfast increase in computational resources has enabled cardiac models of unprecedented complexity and realism. While the first electrophysiological model described the properties of single cardiac myocytes with a simplified set of equations, it is nowadays possible to connect millions of biophysically detailed models of cardiac electrophysiology such that they characterize the bioelectric phenomena in slabs of tissue or even the whole heart.

The electrical activity of the heart creates electrical fields which finally lead to potential differences on the body surface. These potential differences can also be calculated by solving the so-called forward problem of electrocardiography. This closes the loop of multiscale modeling as all relevant microscopic (i.e. ion channel gating on the cellular levels) and macroscopic processes (i.e. action potential propagation and solution of the forward problem) are covered by adequate in-silico models.

There are numerous application areas which are predestined for realistic multiscale models. Among them is the prediction of the effects of drugs or genetic mutations on the ECG as well as investigations in the realm of ischemia. At present, these models can help to advance our knowledge on pathological processes or help to assess the plausibility of theories (e.g. with respect to the triggers of arrhythmic episodes, or the connection between dispersion of repolarization and T-Wave morphology). Yet in the future, multiscale models will have the potential to enable patient-specific diagnosis and therapy which is hoped to increase the success of treatment and reduce the costs of follow-up examinations.

1.2 Focus of the Thesis

This thesis is focused on several different aspects within the whole loop of multiscale modeling. Both microscopic and macroscopic topics have been investigated. On a cellular level:

- The inclusion of a model for beta-adrenergic signaling has enabled the consideration of sympathetic effects on cellular electrophysiology.
- The modeling of the congenital Long-QT syndrome has provided an interesting example to investigate the effects of a genetic mutation both on a cellular scale and on the body surface ECG.

On an organ level:

- A model to represent the effects of the excitation conduction system (in particular the Purkinje fiber network) was developed as such a model is vital for the realistic simulation of ventricular activation.
- The effects of various different distributions of electrophysiological heterogeneities were investigated with respect to their role in the genesis of the T-Wave.

On the torso level (solution of the forward problem of electrocardiography):

- Different rule-based approaches were tested to model the distribution and orientation of skeletal muscle fibers.
- A dynamic model of a deforming heart was developed to assess the effects of contraction and relaxation on the morphology of the T-Wave.
- The influence of different tissues was ranked with respect to a realistic solution of the forward problem of electrocardiography.
- A new method was developed to predict the effects of conductivity variations in important organs during the solution of the forward problem.

1.3 Structure of the Thesis

Part I covers basic medical and technical principles:

- **Chapter 2** provides the necessary medical foundations (anatomical basics, electrophysiological basics, electrophysiological heterogeneities, the excitation conduction system, beta-adrenergic regulation, the congenital Long-QT syndrome and the electrocardiogram).
- **Chapter 3** presents the state of the art in cardiac modeling. Yet this overview is not exhaustive as only modeling areas are covered which were also in the focus of this thesis.

Part II outlines the methodology that was used in each of the presented studies:

- **Chapter 4** introduces the methods that were used for all studies within the realm of electrophysiological modeling; i.e.:
  Modeling of the Specialized Conduction System: Parts of this work were created during the supervised Student Research Project of Raffi Kalayciyan [1]. Some of the corresponding results were also presented at a scientific conference [2, 3].
  Modeling of Electrophysiological Heterogeneities: Parts of this work were submitted to a scientific journal [4].
  Modeling of Beta-Adrenergic Regulation: Parts of this work were created during the supervised Diploma Thesis of Carola Otto [5] and the Student Research Project of Andreas Bohn [6]. Parts of the results were also presented at a scientific conference [7].
  Modeling of the Congenital Long-QT Syndrome: Parts of this work were created during the supervised Student Research Project of Andreas Bohn [6]. Parts of the results were also presented at three scientific conferences [8, 9, 10].
- Likewise, **Chapter 5** covers the methods used for all anatomical studies and all investigations based on the solution of the forward problem of electrocardiography; i.e.:
  Anatomical Modeling: With respect to the anatomical modeling no new methods were developed, but the already existing software from the Institute of Biomedical Engineering (Karlsruhe Institute of Technology) was merely used.
  Resolution Effects of Anatomical Models: Parts of the results were presented at a scientific conference [11].
  Modeling of Skeletal Muscle Fiber Orientation: Parts of the results were presented at a scientific conference [12].

Modeling of the Effects of Ventricular Deformation on the ECG: Parts of the results were presented in a scientific journal with contributions from Oussama Jarrousse and Thomas Fritz [13].

Ranking the Influence of Various Tissue Conductivities: Parts of this work were created during the supervised Master Thesis of Pham Tri Dung [14]. Parts of the results were also presented in a scientific journal (in collaboration with Frank Weber) [15].

Predicting the Effects of Tissue Conductivity Variations Based on the Principal Component Analysis: Parts of this work were created during the supervised Diploma Thesis of Stefan Bauer [16]. Parts of the results were also presented on a scientific conference [17] and in a scientific journal (in collaboration with Frank Weber) [18].

Part III presents the results of the studies that were introduced in **Part II**:

- **Chapter 6** provides the results of all electrophysiological studies. The methods used in each of these studies are described in **Chapter 4**.
- Likewise, **Chapter 7** presents the results of all anatomical studies and studies that were related to the solution of the forward problem of electrocardiography. The methods used in each of these studies are described in **Chapter 5**.

The thesis concludes with a summary and an outlook on possible future work in **Chapter 8**.

Part I

Basic Foundations

2

Medical Background

This chapter is intended to provide the medical foundations which are necessary for most of the technical explanations following hereafter. The description of the medical background was restricted to the essential minimum. However, most of the sections conclude with references on more specialized literature to which the interested reader is kindly referred.

2.1 Cardiac Anatomy

2.1.1 Anatomical Overview

The pumping action of the heart drives the circulatory system, thereby sustaining a constant blood flow, which distributes oxygen to the organs in the periphery and removes metabolic wastes. During one cycle, deoxygenated blood from the peripheral organs enters the right atrium through the superior or inferior vena cava. From there, it is pumped to the right ventricle and into the lungs where it is oxygenated again. After that, the blood flows through one of the four pulmonary veins into the left atrium. From the left atrium the blood is pumped into the left ventricle and subsequently into the aorta from where it is distributed throughout the body. From Fig. 2.1 it can be seen, that the left ventricular wall is much thicker than the wall in the right ventricle or the atria. This is due to the pressure difference between the systemic loop ($\approx$ 80-120 mmHg) and the pulmonary loop ($\approx$ 30 mmHg).

The uni-directional flow of the blood is enforced by a number of different valves between atria and ventricles and between the cavities and the arteries/veins. The mitral and tricuspid valve which seal the atria from the ventricles are connected to the papillary muscles via the chordae tendineae.

Other important anatomical aspects like the ventricular fiber orientation or the specialized excitation conduction system are introduced in subsequent sections

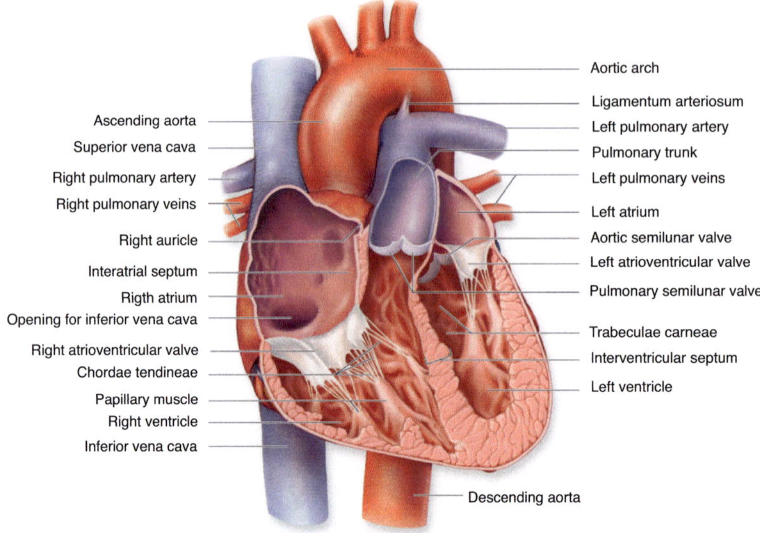

Fig. 2.1. The human heart has four cavities which are separated into a left and right side by the septum. A normal heart weighs between 230 and 350 g. Most of the weight is contributed by the left ventricle which has a much ticker wall than the right ventricle. Figure modified based on [19].

(see 2.1.2 and 2.2.3). A complete anatomical overview introducing all important structures of the heart is outside the scope of this thesis. For more information, please refer to [20, 21, 22].

2.1.2 Ventricular Fiber Orientation

Ventricular fiber orientation is one of the key factors that influence the activation sequence of the ventricles. This is due to the fact that conduction velocity is anisotropic with faster propagation along the fibers than perpendicular to them [23, 24].

There are a number of different experimental techniques that have been used to characterize ventricular fiber orientation in the past. They range from quantitative polarized light microscopy [25] over different histological sectioning techniques [26, 27] to Diffusion Tensor MRI (DTMRI), which has been originally developed to track nerve fibers in the brain but has recently also been used increasingly often to image fiber orientation in various ventricular preparations [28, 29, 30, 31, 32, 33, 34].

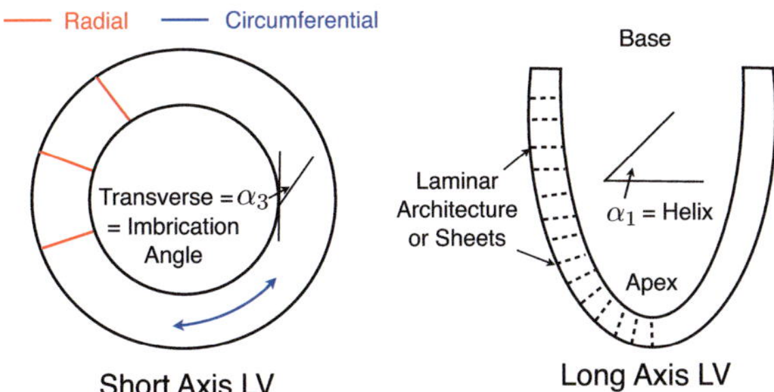

Fig. 2.2. Schematic drawing, clarifying the terminology that is used to describe the ventricular fiber orientation. Usually, a local coordinate system is introduced with respect to which the fiber angles are determined. This local coordinate system can be formed by the longitudinal axis of the left ventricle, the vector which is perpendicular to the epicardium and a third vector which can be calculated by their cross product. In this local coordinate system, the helix angle α_1 describes the fiber inclination in apico-basal direction while the transverse angle α_3 characterizes the degree to which the fibers imbricate. A non-zero transverse angle means that the fibers are not parallel to the endo or epicardial surface. Finally, the schematic drawing indicates the arrangement of sheets that have been reported to run in radial direction [39].

Most studies that investigated the fiber orientation quantitatively introduced a local coordinate system with respect to which the fiber angles were determined. Care has to be taken though if the results from different studies are compared as there are differences between the local coordinate systems (e.g [35] vs. [29, 30, 34, 36]. Fig. 2.2 introduces the most common terminology used to describe the fiber arrangement: in this context, the helix angle α_1 describes the fiber inclination in apico-basal direction whereas the transverse angle α_3 characterizes the degree to which the fibers imbricate (i.e. are not parallel to the endo or epicardial surface). The remaining angle α_2 is usually not used to characterize the course of the fiber orientation as changes in it can not be easily interpreted (as e.g. in case of α_1 or α_3). It can be calculated from the other angles as described in [37, 38].

Concerning the course of the helix angle α_1 it is commonly accepted that it rotates more or less linearly through the wall (from endo to epicardium as can be seen in Fig. 2.3; this is also called "fiber twist"), e.g [35, 26, 40, 41, 30, 25, 37]. Only the degree of rotation and the question of variability between different regions of the heart is discussed [26].

In case of the transverse angle α_3 the situation is more complicated. In a histological study, the transverse angle was reported to be close to 0 $^\circ$ (which means that the fibers have a circumferential orientation) with minimal variation from apex

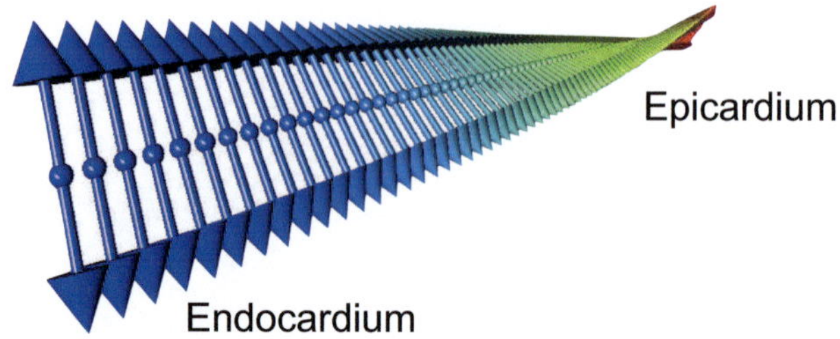

Fig. 2.3. Transmurally rotating helix angle. Blue color represents a positive helix angle whereas red denotes negative angles.

to base ($\alpha_{3,Apex} = -3°$ vs. $\alpha_{3,Base} = 3°$ according to [35]). This is not surprising, as histological studies have to rely on pre-defined cut surfaces. While this allows to measure the in-plane component (helix angle) with high accuracy, possible out-of-plane components (transverse angle) remain unknown [29]. In this case, DTMRI has significant advantages, as it allows to measure the fiber orientation non-destructively with a dense spatial sampling. Several DTMRI based studies have reported a significant apico-basal variation of the averaged or midwall transverse angle [29, 34, 30]. This could also have important elastomechanical implications [36, 41]. Moreover, a possible transmural variation of the transverse angle has been suggested [42, 43, 37, 27]. Although this is still debated, as fibers are generally assumed to be parallel at the endo and epicardial surface, a combination of myocyte branching [44] and volume averaging effects within each voxel could deliver a potential explanation.

Finally, fibers have been reported to be arranged in a laminar fashion with extensive cleavage planes, that are 4-6 myocytes thick [39]. These planes (or sheets) run radially from endo to epicardium (see Fig. 2.2). Distinct mechanical and electrical properties have been associated with this laminar architecture [45]. However, it should also be noted that the cleavage planes might be artifacts that are due to tissue shrinkage during the preparation process and that there existence is doubted by Lunkenheimer *et al.* [27].

A more detailed overview comparing the results of different studies that investigated the fiber orientation in the ventricles can be found in [37].

2.2 Cardiac Electrophysiology

2.2.1 Electrophysiological Basics

Cardiac cells are confined from the surrounding extracellular space by a selectively-permeable membrane. Under physiological conditions, there are different ionic concentrations in the intra- and extracellular space contributing to a potential difference that can be measured over the cell membrane and which is referred to as transmembrane voltage. However, as cardiac myocytes are electrically excitable, this transmembrane voltage is usually not constant but changes in a controlled fashion that is determined by various ion channels, pumps and exchangers which are integrated into the cell membrane.

If the transmembrane voltage rises from its resting voltage (usually around -90 mV) to more positive values, an action potential (AP) is triggered (see Fig. 2.4). Phase 1 of each ventricular action potential is determined by a steep increase in sodium channel conductivity g_{Na}. The inflow of positively charged sodium raises the transmembrane voltage further, creating a positive feedback. This causes the remaining sodium channels to open and thus leads to a rapid depolarization. After several milliseconds, the sodium channels close again and phase 1 of the AP is stopped. During the plateau phase (phase 2), the calcium conductivity g_{CaL} is increased. The following inflow of calcium triggers an intracellular calcium release from specialized storage compartments. This calcium release is responsible for the initiation of mechanical contraction (for details on this so-called excitation contraction coupling, please refer to [46]). The repolarization of the myocyte (phase 3) is determined by the opening of voltage gated potassium channels (increase of g_K). The efflux of the positively charged potassium leads to a restoration of the transmembrane voltage to its negative resting values (phase 4). Usually, the sodium channels can not be reactivated until the repolarization is almost completed. This is called refractoriness of a cell and is one of many safety features that protect the heart from (high frequency) arrhythmic events.

As explained above, electrical excitation is an important trigger, as it precedes mechanical contraction. To maximize the pump function of the heart, it is essential that all cells contract in a coordinated fashion. This is partly ensured by the coupling of the myocytes through so-called gap junctions. Thus the heart is an electrical syncytium, which means that all cells will be excited if an excitation is triggered somewhere in the heart.

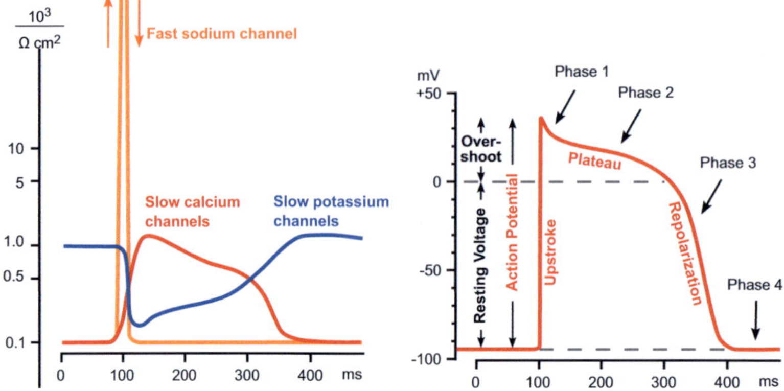

Fig. 2.4. Conductivity changes of the most important ion channels during the course of an action potential can be seen on the left side. On the right side, the morphology of an action potential is displayed. Its different phases are determined by the controlled influx and efflux of the ionic currents depicted on the left side. Figure modified based on [47].

2.2.2 Electrophysiological Heterogeneities and Dispersion of Repolarization

It is commonly accepted that some of the ion channels that are responsible for the formation of the ventricular action potential have heterogeneous densities or biophysical properties throughout the cardiac tissue. Such heterogeneous electro-physiological properties have been reported for I_{to} [48, 49, 50, 51, 52, 53, 54, 55, 56, 57], I_{K1} [53, 58, 59], I_{Ks} (a detailed overview is provided in section 4.2.3), I_{NaCa} [60], $I_{Na,L}$ [61] and I_{up} (also called SERCA) [62]. They are hold responsible for a differentiated reaction of the myocytes towards pharmacological agents [63] as well as the synchronization of contraction across the ventricular wall [64, 65]. Furthermore changes in the physiological configuration of heterogeneities can facilitate the formation of a substrate that could trigger arrhythmogenic events [63]. Apart from that, electrophysiological heterogeneities (in humans particularly heterogeneities in I_{Ks} [66]) cause differences in action potential duration (APD) in myocytes from different regions of the ventricles. These APD differences lead to dispersion of repolarization (DOR), which is commonly considered to be the reason for T-Wave concordance (i.e. R-Peak and T-Wave have the same polarity). Different types of dispersion have been reported: transmural dispersion (through the wall of the ventricles), apico-basal dispersion (from apex to base) and interventricular dispersion (from left to right ventricle) [67]. Yet it is still controversially discussed to which extent these dispersion contribute to the genesis of the T-Wave and which of them is predominant [68].

A lot of experimental studies in the past tried to link APD dispersion and T-Wave shape and derived rules for the most likely sequence of repolarization. Some of them were based on the idea that heat shortens the APD while cooling prolongs it. The associated changes of T-Wave shape and polarity were subsequently observed and interpreted. Such experiments were conducted on a number of different species. As not all species have positive T-Waves under baseline conditions the results of these studies have to be carefully interpreted:

- Frogs normally have negative T-Waves. However, the T-Wave became positive after a warming of the apex shortened apical APD.
- Turtles have positive T-Waves under baseline conditions. In this case, a cooling of the apex prolonged apical APD and led to inverted T-Waves.

The results of these and similar experiments are summarized in [69].

2.2.3 The Specialized Excitation Conduction System

The excitation conduction system of the heart consists of specialized cells which control and coordinate both the generation and conduction of action potentials throughout the myocardium. Thus it can ensure the appropriate rate and timing of the contraction in all regions of the heart, which is essential for the effective cardiac function. Unlike normal atrial or ventricular myocytes, the cells of the excitation conduction system have the ability to depolarize spontaneously and thus determine the heart rate. Under physiological conditions, the sinus node (SN) has the highest autorhythmicity. From there, the depolarization propagates over the Bachmann's bundle to the left atrium and across the terminal crest and the pectinate muscles into the right atrium as visualized in Fig. 2.5A (for anatomical details on fast conducting structures in the atria, please refer to [70]). Once it reaches the atrioventricular node (AVN), the conduction is delayed so that atrial contraction occurs before the contraction of the ventricles (thereby atrial contraction can help to fill the ventricles with blood). From the AVN the depolarization wave travels through the bundle of His and is split up into the so-called right bundle branch (RBB) and left bundle branch (LBB), which conduct the excitation into the left and right ventricle, respectively (see Fig. 2.5B). Both RBB and LBB are finally connected to the Purkinje fiber network. This Purkinje fiber network has a web-like structure that fans out over the endocardial surface of both ventricles (some fiber even run through the lumen of the ventricles and are thus called free-running Purkinje fibers). Exemplary visualizations of the Purkinje fiber network for different species can be seen in Fig. 2.6.

Part of the Purkinje network are the so-called Purkinje muscle junctions (PMJs), which form the only connections of the excitation conduction system with the myocardium (all previous structures in the ventricles like RBB, LBB and major parts of the Purkinje network are electrically isolated from the working myocardium).

The cells of the excitation conduction system are also known to have unique electrophysiological characteristics. E.g. the Purkinje fiber cells differ from normal ventricular myocytes with respect to the following features:

- There are differences in the magnitude of some ionic currents [76, 77, 78] (e.g. I_{CaL} is smaller [78], I_{K1} is smaller [77, 78], I_{to} is larger [77] in Purkinje cells whereas I_f only exists in Purkinje but not in ventricular cells [78]).
- dV_{max}/dt is greater in Purkinje cells (400-800 V/s) compared to ventricular tissue (150-300 V/s) [78].
- The AP plateau is lower in Purkinje cells [78, 79] probably due to differences in intracellular calcium handling.
- The APD restitution slope was also reported to be different [78, 79].
- Purkinje cells have a different response to pharmacological agents [79], thus they can form a substrate that facilitates the development of early after depolarizations (EADs) which could trigger episodes of Torsade de Pointes (TdP) [78].

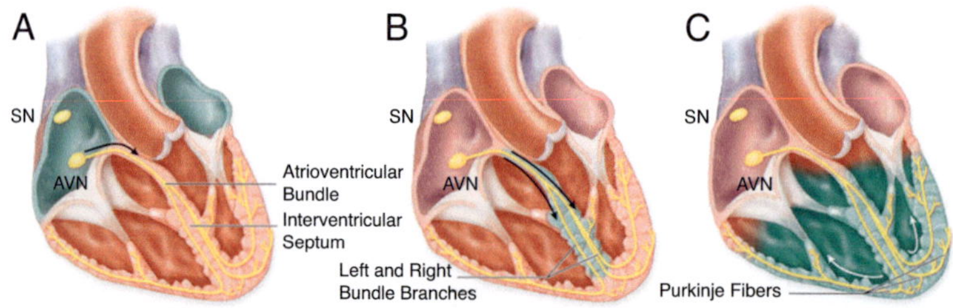

Fig. 2.5. Schematic drawing of the excitation conduction system. A: Depolarization is initiated in the SN and travels from there across the atria to the AVN. B: After a delay, the depolarization front is split up into the RBB and LBB which guide the excitation front into the right and left ventricle, respectively. C: RBB and LBB are connected to the Purkinje fiber network which has a fanlike structure and covers the ventricular endocardium almost completely. Finally, the depolarization spreads into the working myocardium at the so-called PMJs. Figure modified based on [19].

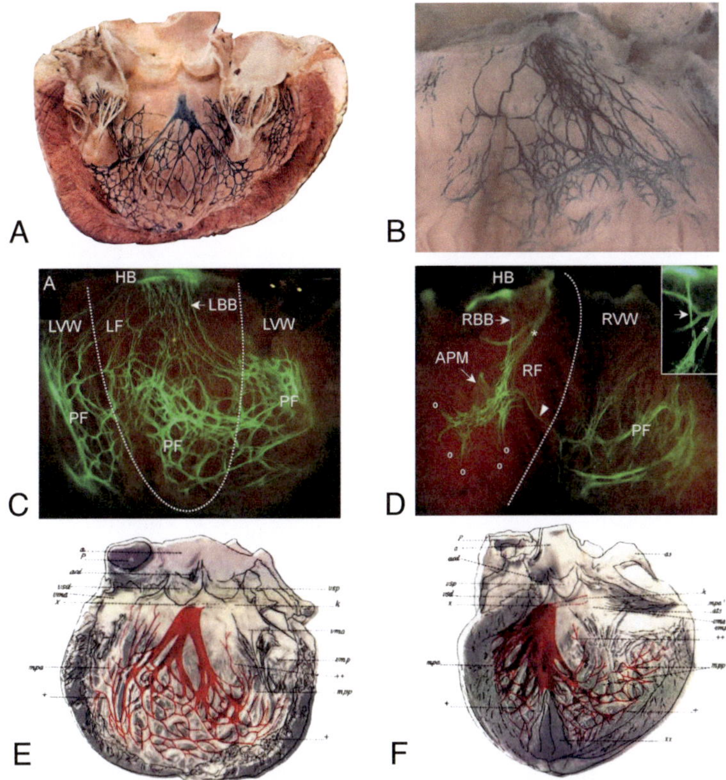

Fig. 2.6. Overview over the ventricular part of the excitation conduction system in different species. A: Fibers of the excitation conduction system in a bovine heart stained with India ink (figure was adapted from [71]). B: Septal surface of the left ventricle in an adult murine heart (figure was adapted from [72]). C: "The left ventricular free wall (LVW) was incised in the center from base to apex, then the two parts of the wall were pulled back on both sides to expose the left flank (LF) of the interventricular septum. The dotted line indicates the limits between the LF and the LVW". Lower density of the Purkinje fibers were found in the right compared to the left ventricle (HB: his bundle, PF: Purkinje fiber). Purkinje fibers were visualized with enhanced Green Fluorescent Protein (EGFP). (figure was adapted from [73]). D: "The whole right ventricular wall (RVW) was pulled back on the right. The dotted line indicates the limits between the right flank (RF) of the interventricular septum and the RVW. Arrowhead indicates a fiber connecting the RF web to the RVW network. Small white circles indicate connecting fibers which have been cut. Insert shows details of the RBB (arrow) which emerged from the His bundle and intersected with the septal artery (star) and its ramifications" (APM: anterior papillary muscle; figure was adapted from [73]). E: Branching of the LBB and the following Purkinje fiber network (figure was adapted from [74] where it has been scanned from Tawara's drawings [75]). RBB and the following Purkinje fiber network (figure was adapted from [74] where it has been scanned from Tawara's drawings [75]).

Yet great care has to be taken when comparing the results of studies conducted on different species as some electrophysiological features are highly species-dependent [80].

In addition to that, anatomical features of the Purkinje fiber network have been reported to vary between different species as well [81, 82]. E.g. the bovine heart has intramural Purkinje fibers that extent through the wall from endocardium almost up to the epicardium (i.e. they stop in the subepicardium, 2 mm before the epicardial surface). This explains why the bovine heart has similar QRS complex durations as a human heart although it is roughly four times as large. Finally even within the same species, there are divergent descriptions of the internal organization and topographical location of the bundle branches and the Purkinje fibers [83]. In human, the LBB is often reported to be divided into two branches called the left anterior and left posterior fascicles [84, 85]. These fascicles are often directed towards the bases of the anterior and posterior papillary muscles of the left ventricle. However, there are other findings in which three rather than two separate divisions were found [86] and studies which found a "diffuse fanlike structure broadly distributed over the left septal surface" [83]. On top of that, it seems that there is also considerable inter-individual variation [86] which makes it difficult to derive generally applicable rules with respect to the anatomy of the human excitation conduction system.

2.2.4 Beta-Adrenergic Regulation of Cellular Electrophysiology

The autonomic nervous system (ANS) subconsciously regulates the activity of a large number of organs in the human body. In situations of stress and physical activity it adapts the heart rate, controls the blood pressure and prioritizes the blood supply of vital components such as the brain and the muscles while other processes are suppressed (e.g. digestion). The ANS is classically divided into two subsystems: the parasympathetic system and the sympathetic system (see Fig. 2.7). Both systems usually work antagonistically with respect to each other.

In this thesis, the focus is on the sympathetic regulation of the heart. In this case, the so-called β_1 receptor is the predominant target for the regulatory hormone Norepinephrine (NE). A similar effect as NE is attributed to the substance Isoproterenol (ISO), which is a sympathomimetic drug that mimics the effects of agents of the sympathetic nervous system. It is therefore often used in electrophysiological experiments to simulate the influence of an active sympathetic regulation.

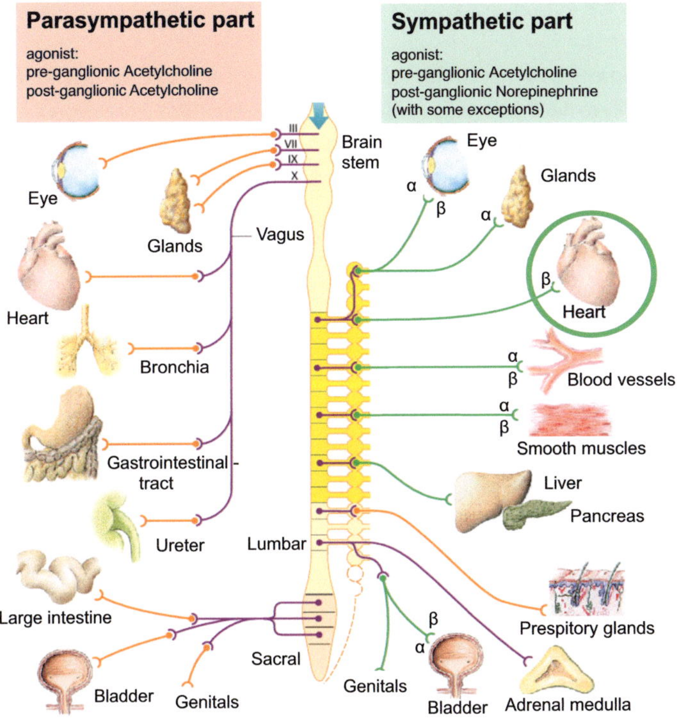

Fig. 2.7. Schematic drawing detailing the innervation of the various organs by the parasympathetic and the sympathetic branch of the autonomic nervous system. Various different hormones and receptors are involved in this regulatory system. However, when it comes to the heart, β_1 receptors are predominantly responsible for the sympathetic regulation. Figure was modified based on [87].

If activated, the sympathetic branch of the ANS influences both cells from the specialized conduction system and cells from the ventricular and atrial working myocardium. In case of the sinus node, the sympathetic influence has chronotropic effects (increase in heart rate). Similarly, beta-adrenergic regulation induces positive dromotropic effects in the atrioventricular node (increase in conduction speed). Regarding the working myocardium, an increased activity of the sympathetic branch of the ANS increases the inotropy (larger contractility) and the lusitropy (faster relaxation of the heart).

Further details on the different types of receptors and hormones as well on the pre- and postganglionic processes that are participating in the regulatory control of the ANS are outside the scope of this thesis. For more information, please refer to e.g. [88, 87].

A description of the intracellular signaling cascade that is activated by the binding of NE or ISO to the β_1 receptors will be provided in section 4.3. There, we will also introduce an in-silico model, which is able to quantitatively describe the effects of this adrenergic signaling cascade on a number of different ion channels.

2.2.5 The Congenital Long-QT Syndrome

General Introduction

The congenital Long-QT Syndrome (LQTS) denotes various ion channel mutations with a prevalence of $\approx$ 1:5000 [89, 90] (however not all mutation carriers are symptomatic [91]). Being "little more than a medical oddity" at first with "a minute number of investigators interested", the exploration of the LQTS benefited tremendously from the foundation of the International Long-QT Syndrome Registry in 1979 [92]. By providing a repository for genotyped families (in 2003 over 1200 LQTS families were available) and a platform for data exchange and discussions, the registry has enabled many of the ground-breaking discoveries within the realm of the LQTS that have been made ever since [92].

Patients suffering from LQTS often experience polymorphic ventricular tachycardia from the Torsade de Pointes (TdP) type which can lead to sudden cardiac death if untreated. In the clinical practice, LQTS patients are usually identified by a prolonged QT_c time ($QT_c > 0.46$ s indicates a potential LQTS [93]). However, if diagnosis is merely based on QT_c prolongation it is not very specific as not all LQTS patients show such a prolongation. In an effort to enhance diagnostic reliability, Schwartz *et al.* [94] thus introduced a more complex point score system that considers other factors such as the occurrence of syncopes or the family history.

It is important to consider that there is not a single LQTS but rather there are different LQT subtypes (genotypes) that have been discovered in recent years (e.g. in 1998 four subtypes were known [95], in 2003 already seven subtypes were discovered [96] and today the list of LQT subtypes encompasses 12 entries [97]). Subtypes are grouped depending on the gene which is affected by the mutation. Table 2.1 gives an overview over the various types of LQTS including genes, proteins, ion channel currents and effects of the various mutations.

It is evident from Table 2.1 that LQT 1-3 are the most common subtypes:

- In case of LQT 1, the slow delayed rectifier potassium current I_{Ks} is affected. The mutation causes a loss of function (reduction of the maximum channel current).

- In case of LQT 2, the rapid delayed rectifying current I_{Kr} is reduced (here the mutation causes a loss of function as well).

- Concerning LQT 3, the mutation leads to a re-activation of the sodium current (late I_{Na}, gain of function).

The standard procedure to determine the LQT subtype is genetic screening. This is both labor intensive and costly [101]. Yet it is imperative that the patient's subtype is known as therapy (see below) and TdP triggers are different for each of the three most important subtypes.

Classically, patients suffering from LQT 1 are thought to have the greatest risk of cardiac events during exercise (e.g. swimming), conditions of elevated heart rate, and sympathetic nerve activity. In contrast to that, cardiac events in LQT 2 patients were triggered by arousal/emotions or noise (e.g. alarm clock). Finally, LQT 3 patients have the greatest risk of TdP during periods of rest, bradycardia and conditions of low sympathetic nerve activity (a summary of these behavioral triggers can be found in [102]).

Table 2.1. List of LQT subtypes linking affected genes, proteins, ion channel currents and effects of the various mutations. Data for the table was compiled from [98, 97, 96]. The sources did not always agree as to the genes and affected ionic currents. Furthermore, some publications associate a mutation in the A-kinase anchoring protein with LQT 1 [99, 100] which is in contradiction with the publication from Zareba *et al.* [97] where it is referred to as LQT 11 (as listed in this table).

Subtype	Gene	Protein	Current/Effect	Occurrence
LQT 1	KCNQ1/KVLQT1	α subunit	$I_{Ks}\downarrow$	30-35%
LQT 2	KCNH2/HERG	α subunit	$I_{Kr}\downarrow$	25-30%
LQT 3	SCN5A	α subunit	Late $I_{Na}\uparrow$	5-10%
LQT 4	ANKB	ankyrin-β	$I_{Na}*$	$< 1\%$
LQT 5	KCNE1/minK	β subunit	$I_{Ks}\downarrow$	$< 1\%$
LQT 6	KCNE2/MiRP1	β subunit	$I_{Kr}\downarrow$	$< 1\%$
LQT 7	KCNJ2	α subunit	$I_{K1}\downarrow$	$< 1\%$
LQT 8	CACNA1	α subunit	$I_{CaL}\uparrow$	$< 1\%$
LQT 9	CAV3	caveolin 3	$I_{Na}*$	$< 1\%$
LQT 10	SCN4B	β subunit	Late $I_{Na}\uparrow$	$< 1\%$
LQT 11	AKAP9	A-kinase anchoring protein	$I_{Ks}\downarrow *$	$< 1\%$
LQT 12	SNTA1	Sodium current regulator	$I_{Na}\uparrow *$	$< 1\%$

Review articles characterizing the LQTS are available in [98, 97, 96, 103].

Adrenergic Influence on the Long-QT Syndrome

It has long been known that some LQT subtypes react highly sensitive to the presence of beta-adrenergic agents. An imbalance of the various sympathetic inputs was initially even hold responsible for the existence of the syndrome [104] (before the underlying mutation of ion channel proteins was discovered).

As explained above, TdP in some LQT subtypes is thought to be triggered by sympathetic activity. This was confirmed by Noda et al. [105] who found that the change in ventricular repolarization that was induced by sympathetic activity was different between LQT 1-3. Similarly, Schwartz *et al.* reinforced the role of autonomic response in patients suffering from LQT 1 [106].

Although the link between sympathetic activity and TdP in LQTS is clinically well established (see above), the cellular mechanisms which could explain these differentiated responses are still not completely understood (see paragraph "Open Questions" below). This is different with respect to a relatively new mutation, which is known to selectively disrupt the sympathetic regulation of the slow delayed rectifier potassium current I_{Ks} [107, 99, 108, 100, 109]. In this case, the adaptor protein Yotiao is affected by the mutation (AKAP-binding domain mutation). Thus, I_{Ks} cannot be regulated by the sympathetic nervous system yet the regulation of the remaining channels is still intact. Under increased sympathetic activity, this means that the increase in I_{CaL} cannot be compensated by an increase in I_{Ks}. The resulting APD and QT prolongation is here truly restricted to periods of physical activity.

A summary of the sympathetic modulation of the LQTS (not considering recent findings like the AKAP-binding domain mutation) can be found in [110, 111].

The Use of the Electrocardiogram in Long-QT Diagnosis

For a general introduction of the electrocardiogram, please refer to section 2.3. Electrocardiographic studies on the LQTS can be separated based on whether they tried to differentiate between LQTS patients and healthy control subjects or whether they tried to discriminate between the various LQT subtypes. Studies from the first group used principal component analysis [112] or body surface potential mapping [113]. However both methods did not have an impact on the present clinical practice which is still based on the standard 12 lead ECG and the point score system from Schwartz *et al.* [94]. Studies from the second group aim at providing

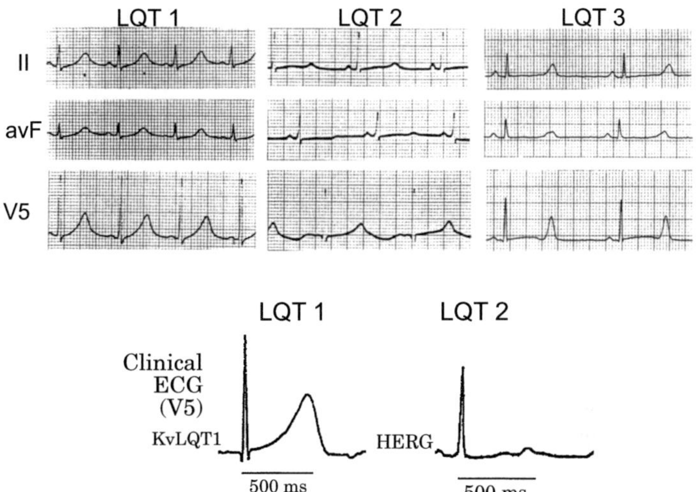

Fig. 2.8. Representative examples for ECGs that are associated with patients suffering from LQT 1-3. T-Wave morphology differences are evident between the three mutations: While LQT 1 patients have broad based T-Waves, patients suffering from LQT 2 have small, notched T-Waves. Finally, the ECG of LQT 3 is associated with an isoelectric ST segment and a narrow T-Wave. ECGs in the top row were adopted from [118] whereas the ECGs in the bottom row originate from [119].

a good first initial guess which can be used to direct the genetic screening towards a specific mutation.

In general, the following ECG features are attributed to the three most common types of LQTS [114, 115]:

- LQT 1 patients often show broad based T-Waves of relatively high amplitude (T-Wave amplitude LQT 1: 0.34 ± 0.16 mV vs. LQT 2: 0.21 ± 0.25 mV vs LQT 3: 0.30 ± 0.38 mV [114]).
- The T-Waves in LQT 2 are of low-amplitude with a bifid or notched [116] morphology. Typical morphological features and differences compared to LQT 1 can even be amplified by subjecting the patients to a stress test [117].
- Finally, the ECG in LQT 3 shows a delayed, narrow T-Wave and an obvious ST-segment prolongation.

Fig. 2.8 shows examples of typical ECG waveforms for LQT 1-3.

Recently, Struijk *et al.* presented a study in which they used parameters that described T-Wave morphology in terms of duration, asymmetry, flatness and amplitude. With a discriminant analysis based on 4 or 5 parameters, they were able to separate between patients suffering from LQT 1 and LQT 2 [120].

Gender Differences in the Congenital Long-QT Syndrome

Although, LQTS is equally distributed between the sexes, adult women are clinically more often affected by the disease than men (probably due to their longer QT intervals [121, 122]). Concerning younger patients, the probability of a first cardiac event by the age 15 is higher in males than in females [123]. However for male patients this risk decreases after puberty (not for female patients). Finally, gender seems to impact as well on the electrophysiological response to beta-adrenoceptor blockade [124]. Overall it should be noted, that gender and the associated hormonal influences are a major determinant for the course and clinical manifestation of LQTS. A summary of gender-related influences on the LQTS can be found in [103].

Therapeutic Options for Treatment of the Congenital Long-QT Syndrome

Most often, LQT patients that already experienced a syncope are treated with beta-blockers. If syncopes or arrhythmia-related symptoms still occur despite the beta-blocker treatment, the implantation of a cardioverter defibrillator (ICD) is often necessary [125].
It should be noted, that beta-blocker therapy is particularly promising in LQT 1 patients [126] while it is significantly less effective in LQT 2 and might be even pro-arrhythmic in LQT 3 patients (as lower heart rates can provoke cardiac events in LQT 3 patients [96]).
In addition to that, there is also a "genotype-specific" treatment:

- LQT 2 patients can be treated by increasing the extracellular potassium level which increases the repolarizing current I_{Kr} [127]. This current is normally reduced due to the mutation effects.
- LQT 3 patients were reported to benefit from the administration of the sodium channel blocker Mexiletine, which significantly shortened their QT intervals [128].

Finally, therapeutic trials were conducted with left cardiac sympathetic denervation [129]. In this case, the effect of the sympathetic regulation is erased by surgically removing the corresponding ganglia. It should only be considered if beta-blocker treatment was not successful or tolerated by the patient. As it is only available at specialized facilities, it is seldom used in the clinical practice. A more detailed summary of all therapeutic option that exist for the treatment of LQTS can be found in [130, 131].

The Congenital Long-QT Syndrome: Open Questions

Although a multitude of studies concerning the LQTS has been published since its discovery, there are still some unanswered questions:

- As previously mentioned, LQT 1 patients are frequently associated with broad based T-Waves of large amplitude. This is not easily explainable, as the heterogeneously expressed current I_{Ks} which is assumed to be responsible for the dispersion of repolarization is the same current that is reduced in case of LQT 1 (loss of function mutation). One would therefore rather expect a reduced T-Wave amplitude as a reduction of dispersion of repolarization should homogenize the repolarization.

  In case of LQT 2 the situation is vice versa. Here, we would expect a large T-Wave amplitude from theoretical considerations: As the homogeneously expressed current I_{Kr} is reduced, the inherently present heterogeneous properties of I_{Ks} are amplified (due to the fact that the contribution of I_{Kr} to the repolarization is weakened). This should lead to an increased APD gradient and thus to larger and wider T-Waves. Yet clinically, LQT 2 patients often show low amplitude T-Waves (see Fig. 2.8).

  It is possible that adaptation processes (e.g. remodeling) do partly compensate for the effects of the mutations. However no detailed information on these processes is so far available.

- Although patients suffering from LQT 2 often exhibit notched T-Waves, the origin of these notches is still somewhat unclear. In several modeling studies, low extracellular potassium concentrations were reported to be able to induce notched T-Waves [132, 133]. Yet it is questionable if all LQT 2 patients with T-Wave notches have such a low extracellular potassium level.

- Finally it is known from clinical observations, that LQT 3 patients have a high risk of TdP during periods of rest or low heart rate. Antzelevitch *et al.* claim that an increase in transmural dispersion of repolarization is the main trigger for the onset of TdP (e.g [110]). Yet, an increase in sympathetic activity is known to increase the repolarizing current I_{Ks} (see section 4.3.1). Under the assumption that this current is distributed heterogeneously such an increase should lead to an increase in transmural dispersion of repolarization regardless of the type of

LQTS (for LQT 1 and LQT 2 this increase was also observed by Antzelevitch *et al.* but not for LQT 3).

2.3 The Electrocardiogram

Projections of the electrical potential differences on the heart that exist during de- and repolarization can be measured on the body surface. The resulting waveform has been termed the electrocardiogram (ECG) and is still the most important technique when it comes to the non-invasive assessment of cardiac electrophysiology. The typical ECG signal that is associated with a single heart beat can be seen in Fig. 2.9:

- The P-Wave is caused by electrical activation of the atria (typical physiological duration: $\approx 100 - 120$ ms)
- The PQ segment is associated with the conduction delay in the AVN (see section 2.2.3; typical physiological duration: 120-200 ms / typical physiological duration of the PR interval: 113-212 ms [134])
- The QRS complex denotes ventricular activation (typical physiological duration: 69-109 ms [134])
- The ST segment follows the QRS complex and ends with the beginning of the T-Wave (usually the T-onset is not determined and thus no durations for the ST segment are given).
- The T-Wave is a measure for the repolarization sequence of the ventricles. Under physiological conditions it is usually concordant with the QRS complex.
- The U-Wave can sometimes follow the T-Wave. It has a similar shape as the P-Wave. However, its origin is still controversially discussed in the community [135, 136, 137, 138].

In the clinical routine the so-called standard 12-lead recording system is commonly used. It encompasses bipolar (Einthoven) and unipolar leads (Goldberger and Wilson). For a detailed overview concerning different lead systems and their clinical use, please refer to [140]. In general, different leads are sensitive to different wave directions and different regions of the heart (i.e. ischemic changes might not be visible in each/or any of the clinically used leads). The introduction of additional leads can thus help to characterize the electrophysiological processes on the heart more completely.

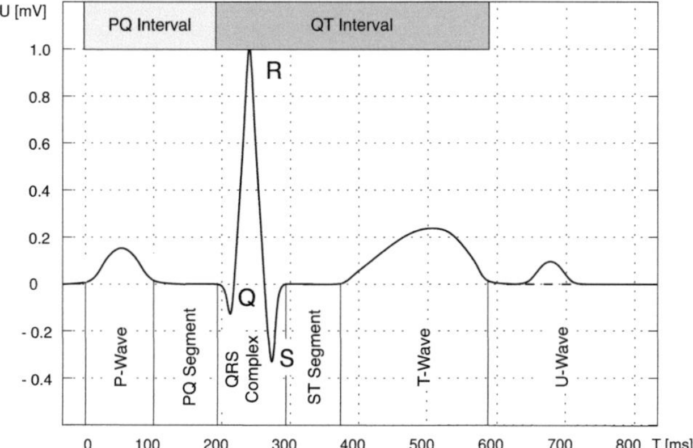

Fig. 2.9. Typical ECG waveform that is associated with a single heart beat. Figure was adapted based on [139].

In the research community the use of multi-channel ECG systems is often favored over the standard 12-lead ECG [141, 142, 143]. In this context, the measured signal is usually called a body surface potential map (BSPM). Several multi-channel ECG recordings have also been performed during the course of this thesis (see section 7.1.1) mainly for validation purposes of the simulation results.

The acquisition system that was used for these measurements (ActiveTwo, BioSemi, Amsterdam, Netherlands) can be seen in Fig. 2.10. All measurements were performed with 64-80 electrodes. For safety reasons, the ECG measurement system was powered by a battery pack and had an optical connection to the laptop which was used for data storage.

To allow a transfer of the electrode locations from the multi-channel ECG measurements to the in-silico model of the torso we used an electromagnetic tracking device (FASTRAK, Polhemus, Burlington VT, USA). In this case, a stylus was used to mark the position of the electrodes manually. The location of the stylus was then saved by the system and used for the electrode mapping procedure.

Fig. 2.11 shows an example of a BSPM measurement on a volunteer together with a screen shot of the retail software (BioSemi) that was used to check the signals during the recording. The acquired signals were post-processed and a template heart beat was created, which could be directly compared to the results of the simulation. More information on the creation of this template heart beat can be found in [146] and in section 4.2.5.

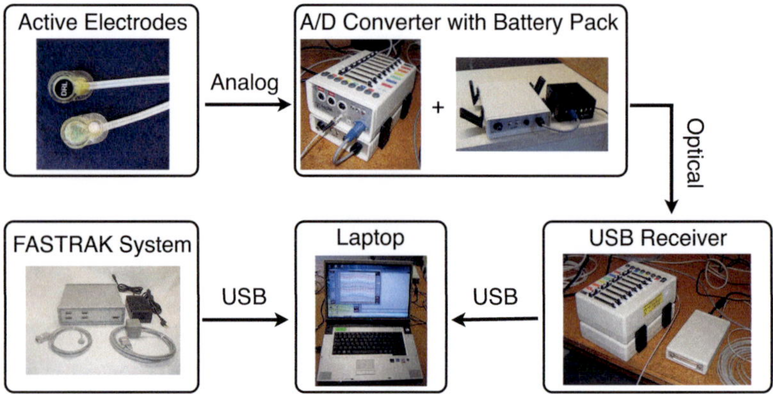

Fig. 2.10. BSPM recording system ActiveTwo from BioSemi. 64-80 active electrodes are used to record the potential differences on the body surface. Each electrode has its own signal amplifier, which increases the signal-to-noise ratio (SNR) thereby enhancing the quality of the recording. The signal is subsequently digitized with 24-bit dynamic range and transferred via an optical and USB interface to a laptop for storage. The electromagnetic tracking system (FASTRAK) uses a stylus to record the position of the electrodes. Data transfer is based on USB as well. Parts of the figure were adapted from Krueger *et al.* (euHeart Project [144]) as well as from [145].

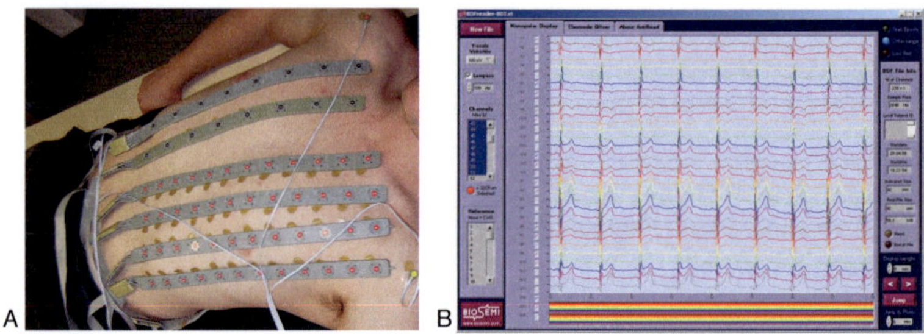

Fig. 2.11. A: Electrode placement (80 electrodes) during the BSPM measurement on a volunteer. The electrode strips facilitate and accelerate the use of the system in a clinical setting. B: Screenshots of the software that was used to save the data and inspect the signal waveforms during the acquisition. Part (B) of the figure was adapted from Krueger *et al.* (euHeart Project [144]).

3

Introduction to Cardiac Modeling

3.1 Modeling Cardiac Electrophysiology

In the following chapter the basic principles of cardiac modeling are introduced. Beginning with the electrophysiological modeling of a single myocyte, the mathematical coupling of these models is described which allows the simulation of bioelectric phenomena in a tissue patch or even a whole heart. After introducing these basic techniques, the state-of-the-art in various fields of cardiac modeling is presented.

3.1.1 Electrophysiological Modeling Basics

Most of the current electrophysiological models are based on the pioneering work of Hodgkin and Huxley [147]. In 1952 they developed the first mathematical model that allowed the quantitative description of cellular electrophysiology. Due to technological reasons their model was based on the giant axon of a squid. They used an electrical equivalent circuit to characterize the ion fluxes through the cell

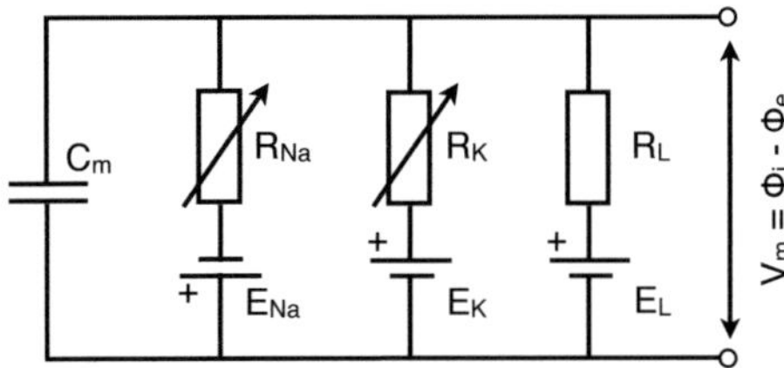

Fig. 3.1. Electrical equivalent circuit created by Hodgkin and Huxley to characterize the electrophysiology of a giant squid axon. The cell membrane was modeled by a capacitance C_m whereas the ion channels were determined by voltage-dependent resistors and the corresponding Nernst voltages were represented by voltage sources.

membrane (see Fig. 3.1). In this case, the membrane was modeled by a capacitance C_m whereas the ion channels were determined by voltage-dependent resistors and the corresponding Nernst voltages were represented by voltage sources. The total transmembrane current I_{mem} consisted out of three components:

$$I_{mem} = I_{Na} + I_K + I_L \tag{3.1}$$

Here, I_{Na} represented the sodium current and I_K the potassium current. All other currents were summarized as leakage currents I_L.

The status of every channel was determined by gates which open and close based on voltage-dependent gating variables (denoted m, n or h). E.g. in case of I_{Na} the corresponding equations were:

$$I_{Na} = g_{Na}(V_m - E_{Na}), \qquad g_{Na} = g_{Na,max} m^3 h \tag{3.2}$$

Here, g_{Na} is the channel conductivity (while $g_{Na,max}$ is the maximal channel conductivity) and E_{Na} is the Nernst potential for sodium. Details on the gating process and the rate constants of the gating variables can be found in [46].

Finally, changes in the transmembrane voltage V_m could be described by:

$$\frac{dV_m}{dt} = -\frac{1}{C_m}(I_{mem} - I_s) \tag{3.3}$$

where I_s was an intercellular stimulus current.

Models of Human Ventricular Electrophysiology by ten Tusscher et al.

The electrophysiological models from ten Tusscher *et al.* represented the state-of-the-art during the course of this thesis. They have several advantages [148] compared to the widely used Luo-Rudy model (which is available in various different versions [149, 150]). Originally developed for the study of reentrant arrhythmias, both ten Tusscher model revisions include distinct parameter sets for endocardial, M and epicardial cells thus allowing the consideration of transmural electrophysiological heterogeneity. The main improvements between the first and second revision were directed towards a more realistic description of the intracellular calcium dynamics (considering a subspace (SS) in the latest version). Fig. 3.2 shows a schematic visualization of both model revisions. Distinct symbols are used to differentiate between the various types of ionic currents and pumps. Furthermore, virtual intracellular ionic compartments are used to group the different ionic fluxes. The total transmembrane current in the first [151] and second revision [152] of the ten Tusscher model contains the following ionic components:

$$I_{mem} = I_{Na} + I_{b,Na} + I_{K1} + I_{to} + I_{Kr} + I_{Ks} + I_{Kp} + I_{CaL} + I_{p,Ca} + I_{b,Ca} + I_{NaCa} + I_{NaK}$$
(3.4)

The currents are:

- *Sodium currents*: I_{Na} = fast sodium influx, $I_{b,Na}$ = background sodium current
- *Potassium currents*: I_{K1} = inward rectifier current, I_{to} = transient outward current, I_{Kr} = rapid delayed rectifier current, I_{Ks} = slow delayed rectifier current, I_{Kp} = plateau potassium current
- *Calcium currents*: I_{CaL} = L-type calcium current, $I_{p,Ca}$ = plateau calcium current, $I_{b,Ca}$ = background calcium current
- *Mixed currents*: I_{NaCa} = sodium / calcium exchanger current, I_{NaK} = sodium / potassium pump current

Recently a new model was published by Grandi *et al.*, characterizing the electrophysiological properties of human ventricular myocytes as well [153]. Among the new model features are updated formulations for the repolarizing potassium currents (I_{Ks} and I_{Kr}) and investigations regarding the effects of a block of these currents on the APD. It might be interesting to repeat some of the investigations presented in this thesis with this model in order to evaluate the effects and plausibility of the new current formulations.

3.1.2 Modeling Studies Evaluating Electrophysiological Heterogeneities

In section 2.2.2 we explained how electrophysiological heterogeneities are responsible for differences in APD that lead to dispersion of repolarization (DOR) and finally determine T-Wave shape and polarity. Although all information concerning these heterogeneous ion channel distributions originate from experimental studies it has to be acknowledged that it is often difficult to directly implicate their findings with specific T-Wave features. Among the reasons for this are low spatial sampling (e.g. in case of electrode recording techniques [154]) or limited field of view (e.g. in case of optical mapping on cut surfaces [155]).
In this context, in-silico models of cardiac repolarization are predestined to fill this gap and overcome the limitations of experimental studies. They can be param-

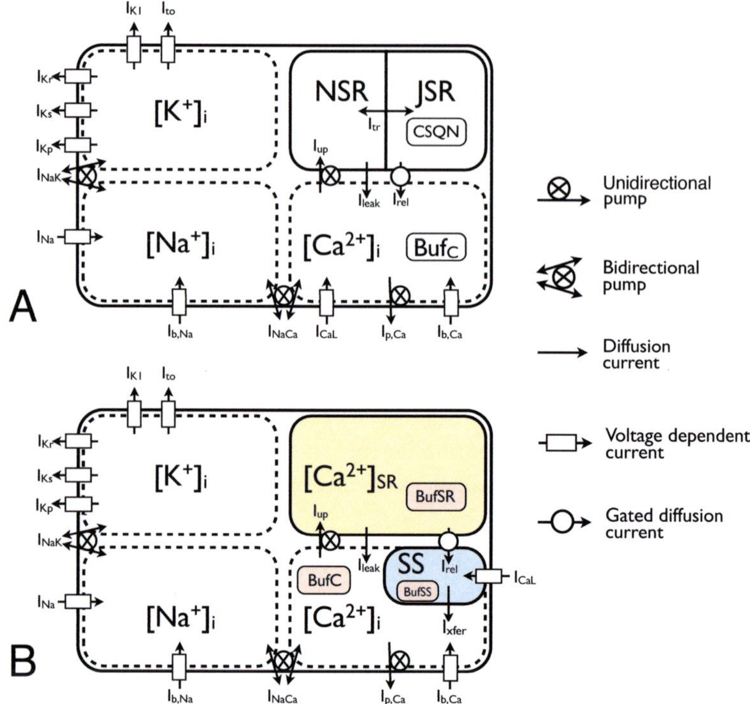

Fig. 3.2. Schematic diagram of the ventricular models from ten Tusscher *et al.* [151, 152]. A: First revision of the ten Tusscher model [151]. The myocyte is divided into several compartments: the cytosol (with the different virtual ionic compartments: K^+, N^+, Ca^{2+}) and the sarcoplasmic reticulum with a calcium up-taking (network sarcoplasmic reticulum: NSR) and a calcium releasing (junctional sarcoplasmic reticulum: JSR) compartment. Between the intracellular compartments, the calcium release current I_{rel} and the calcium uptaking current I_{up} are handling the intracellular flow of calcium. Finally, the buffering of calcium to troponin and calmodulin in the cytosol (Buf_C) and calsequestrin (CSQN) in the JSR are taken into account. B: Second revision of the ten Tusscher model [152]. The main differences compared to the first revision can be found within the intracellular calcium handling. A subspace (SS, also known as dyadic space) was introduced modeling an area where the cell membrane and membrane of the sarcoplasmic reticulum (SR) are in close proximity. Ryanodine receptors I_{rel} are susceptible to an elevation of calcium concentration in the subspace and trigger a release of calcium from the SR. Figure was modified based on [5].

eterized with measurement data and subsequently allow to evaluate the relation between APD dispersion and T-Wave morphology *quantitatively*. This offers significant advantages over the mere *qualitative* interpretations of the past, in which the effects of heterogeneous ion channel properties were usually predicted based on more or less verifiable assumptions. The first study that dealt with quantitative modeling of the T-Wave was from Harumi *et al.* [156]. They presented a model that allowed to predict T-Wave shapes if both activation sequence and APD distribution were known. Subsequent studies used manually adapted dipole sources that characterized the repolarization [157] or measured repolarization times to calculate the corresponding ECG [158].

Due to the continuous increase in computational resources it is nowadays even possible to investigate the influence of a 3D distribution of ion channel heterogeneities on the ECG. This is done by calculating ventricular activation and repolarization in an in-silico model of the ventricles and using the resulting transmembrane voltage distribution as input for the solution of the forward problem (for details see section 3.2.3). Such an approach was chosen by Xue *et al.* [159, 160, 161]. However, their cellular-automaton based model was not able to consider the effects of electrotonic coupling, which severely limits the validity of the presented conclusions [162].

Especially for studies on cardiac repolarization it is known to be imperative to use the more complex and computationally demanding reaction-diffusion models, which can consider the gap-junction related coupling effects (see section 3.1.3). To the best of our knowledge, reaction diffusion models were so far only used to investigate the influence of transmural heterogeneities in a wedge or 1D fiber [163, 133, 164, 165]. Although more complex 3D models are available [166, 167, 168, 169] they were never used to evaluate the effects of apico-basal or interventricular heterogeneities in a systematic way.

3.1.3 Modeling Action Potential Propagation in Tissue

In order to describe the propagation of an electrical excitation wavefront in cardiac tissue, the electrophysiological models that characterize the properties of a single myocyte have to be coupled with a suitable method. Ideally, this coupling method should be able to consider the anisotropic properties of cardiac tissue arising from the non-uniform distribution of gap junctions.

In general, electrical excitation conduction in the heart can be described using *microscopic* or *macroscopic* approaches. In case of the microscopic methods, each myocyte is discretized with a large number of computational elements (μm resolution) and the current flow through the intra- and extracellular spaces can be calculated [170]. An advantage of such an approach is its ability to consider heterogeneities in myocyte shape as well as gap junction, ion channel and capillary distribution. On the downside, when it comes to simulations in larger tissue patches or even the whole heart (with $\approx 10^{10}$ myocytes), the microscopic approaches have to be discarded due to the extremely high computational costs that are involved. In this context, the macroscopic approaches like the bidomain model (here the computational elements have a resolution of several $100\ \mu m$) are preferred which describe the processes of the interacting cells based on a spatially homogenized representation of the tissue [171].

The Bidomain Model

The bidomain model assumes that the cardiac tissue can be described by two domains that represent the volume-averaged intra- and extracellular space. Anisotropy is considered by assigning a unique conductivity tensor for each of the domains. Intra- and extracellular anisotropy originates from the distribution of intracellular structures, gap junctions as well as from the fiber and sheet orientation of the myocardial layers.

If we assume that the potentials in the intra- and extracellular domain are labeled Φ_i and Φ_e, then the corresponding current densities J_i and J_e are determined by:

$$J_i = -\sigma_i \nabla \Phi_i \qquad (3.5)$$

$$J_e = -\sigma_e \nabla \Phi_e \qquad (3.6)$$

In this case, σ_i and σ_e are the volume-averaged conductivities of the intra- and extracellular domain. Furthermore we assume that the principle of charge conservation is applicable here: This means that a current that leaves one domain has to enter the other domain via a transmembrane current I_m:

$$-\nabla \cdot J_i = +I_m(-I_{si}) \qquad (3.7)$$

$$-\nabla \cdot J_e = -I_m(-I_{se}) \qquad (3.8)$$

Here, I_{si} and I_{se} represent potential stimulus currents in the intra- or extracellular domain.

If we now combine equation 3.5-3.8, we receive:

$$\nabla \cdot (\sigma_i \nabla \Phi_i) = +I_m(-I_{si}) \qquad (3.9)$$

$$\nabla \cdot (\sigma_e \nabla \Phi_e) = -I_m(-I_{se}) \qquad (3.10)$$

In a next step, equation 3.10 can be solved for I_m and the result can be used to substitute I_m in equation 3.9:

$$\nabla \cdot (\sigma_i \nabla \Phi_i) = -\nabla \cdot (\sigma_e \nabla \Phi_e) - I_{si} - I_{se} \tag{3.11}$$

Under the assumption that the transmembrane voltage V_m is defined according to $V_m = \Phi_i - \Phi_e$, the left and right side of equation 3.11 can be expanded by adding $-\nabla \cdot (\sigma_i \nabla \Phi_e)$ and thus the equation is transformed into:

$$\nabla \cdot (\sigma_i \nabla V_m) = -\nabla \cdot ((\sigma_i + \sigma_e) \nabla \Phi_e) - I_{si} - I_{se} \tag{3.12}$$

This equation connects the transmembrane voltage with the extracellular potentials and can be used to solve the forward problem of electrocardiography as will be explained in section 3.2.3. It also constitutes the first part of the bidomain model. To derive the second part of the bidomain model which establishes the connection to the electrophysiological model we need the definition of the transmembrane current I_m:

$$I_m = \beta \left(C_m \frac{dV_m}{dt} + I_{mem} \right) \tag{3.13}$$

In this case, β is the membrane surface to cell volume ratio, C_m is the membrane capacitance and I_{mem} is the sum of the ionic transmembrane currents from the electrophysiological model.

Now we can use equation 3.13 to substitute I_m in equation 3.9:

$$\nabla \cdot (\sigma_i \nabla \Phi_i) = \beta \left(C_m \frac{dV_m}{dt} + I_{mem} \right) - I_{si} \tag{3.14}$$

If we replace Φ_i by $V_m + \Phi_e$ then we can transform equation 3.14 and thus acquire the second part of the bidomain model:

$$\nabla \cdot (\sigma_i \nabla V_m) + \nabla \cdot (\sigma_i \nabla \Phi_e) = \beta \left(C_m \frac{dV_m}{dt} + I_{mem} \right) - I_{si} \tag{3.15}$$

The Monodomain Model

The monodomain model is a simplification of the bidomain model and can be derived from the corresponding equations by assuming equal intra- and extracellular anisotropy ratios ($\sigma_i = \kappa \sigma_e$). After applying some algebraic transformations, we finally receive the monodomain equation:

$$\nabla \cdot (\sigma_i \nabla V_m) = (\kappa + 1)\beta \left(C_m \frac{dV_m}{dt} + I_{mem}\right) - I_s \qquad (3.16)$$

in which I_s is the external stimulus current.

Differences between the mono- and bidomain model were reported to be small in case of sinus rhythm and in the absence of extracellularly applied stimuli currents (e.g. defibrillation) [166]. Recent developments in the area of bidomain and monodomain modeling were targeted on a reduction of the computational load of bidomain simulations [172] and on an expansion of the monodomain approach to allow the consideration of bidomain bath-loading effects [173].

As the monodomain equation is computationally cheap compared to a full classical bidomain approach it was used for all large scale biventricular simulations throughout this thesis.

3.1.4 Modeling the Specialized Excitation Conduction System

As explained in section 2.2.3 the specialized excitation conduction system is responsible for the synchronization of ventricular excitation and the coordination of mechanical contraction. Its complex anatomical structure and large inter-species and inter-individual variations (see 2.2.3) pose major challenges concerning its adequate representation in a computer model. In general, three different approaches have been used in the past to model the anatomy and effects of the specialized excitation conduction system:

1) The ventricular tissue was manually stimulated at the sites and corresponding times of early endocardial activation that were reported by Durrer *et al.* [174]. Examples for this approach can be found in [157, 166].

2) The excitation conduction system was created manually based on anatomical atlases or photographic data. Exemplary studies that followed this approach are now listed in chronological order:

- Pollard *et al.* [175, 176] modeled the bundle of His by 35 parallel cables (20 cables formed the LBB while the remaining 15 represented the RBB). These cables branched extensively, but finally merged with 35 reference points on the endocardial surface which represented the PMJs (4 PMJs were located near sites of early endocardial activation while the remaining 31 PMJs were spread throughout the ventricles in an effort to cover large parts of the endocardial surface). In a final step, the PMJs were completely interconnected thus forming the terminal Purkinje network.

- Siregar *et al.* [177] based the architecture of the His-Purkinje system on the data from Durrer *et al.* [174]. The main branches of the specialized conduction system led to three endocardial areas in the left ventricle ("an area high on the paraseptal wall, a central area on the left surface of the interventricular septum, and a posterior paraseptal area at about one third of the distance from base to apex") and one area in the right ventricle (antero-apical region of the endocardium). Electrical activation in the specialized conduction system depended on a segment's length and the associated conduction velocity.

- Berenfeld *et al.* [178] tried to fit the scanned drawings from Tawara *et al.* [75] to the endocardial surface of their ventricular model. The end points of the Purkinje fibers were labeled and used as locations for the PMJs. Ventricular activation was then simulated for physiological or bundle branch block scenarios. In case of unrealistic results, the Purkinje system was iteratively (manually) adapted.

- Simelius *et al.* [179] manually constructed their conduction system based on textbooks on human anatomy and other literature (e.g. the studies from Tawara *et al.* [75] and Durrer *et al.* [174]). The bundle branches were modeled according to the trifascicular concept [85] (e.g. one bundle branch in the right and two branches in the left ventricle). The bundle branches ended at the papillary muscles (both left and right ventricle) and continued as the Purkinje network. In an effort to validate their model, Simelius *et al.* calculated the 12-lead ECG and compared it to ECGs that were acquired on normal subjects.

- Vigmond *et al.* [180] modeled the excitation conduction system by flattening the endocardial surface of both ventricles onto a plane. Then, they manually superimposed a model of the conduction system which was created based on various literature data (for details, see [180]). As there was a high degree of variability between the sources, they concentrated on the following features: "the left Purkinje network had three major areas of activation: the septum, the inferior free wall, and the superior free wall. In the right Purkinje network, the activation proceeded primarily from the septum out to the papillary muscles in the lower part of the ventricle". Finally, the 2D model of the Purkinje network was mapped back on the 3D model prior to the electrophysiological simulations.

- ten Tusscher *et al.* [181] used an interactive graphical software package to place the PMJs and Purkinje fibers on top of the endocardial surface of the left and right ventricle. The initial position of the fibers and PMJs (76 PMJs were placed in the LV while 54 PMJs were positioned in the RV) was based on anatomical textbooks and the sites of earliest endocardial activation reported by Durrer *et al.* [174]. In a subsequent step, the position of the PMJs, Purkinje fibers, fiber length, thickness and conductivities were iteratively (manually) adjusted in order to achieve a realistic activation sequence.

3) The excitation conduction system was created with a semi-automatic procedure that required minimal user interaction. Such a procedure is especially valuable if the excitation conduction system has to be created for a number of different datasets. In this case manual placement of bundle branches and PMJs based on anatomical atlases or photographic data is very time-consuming and thus prohibitive. Studies that use semi-automatic procedures are listed in the following (in chronological order):

- Abboud *et al.* [182] modeled the His-Purkinje conduction system based on a self-similar structure (bifurcating tree). The branches of the utilized bifurcating tree decreased in length with each bifurcation. The angle between the splitting branches was fixed to 70° throughout the whole conduction system. The geometry of the conduction system was thereby determined only by the length of the branches and the splitting angle used.

- Lorange *et al.* [183] used a two-component model to represent the specialized conduction system. In this case, the LBB and RBB were modeled by a system of cables that ended in a sheet of highly conducting tissue which represented the distal Purkinje system. 1120 PMJs were distributed uniformly over the sheets thus connecting it to the remaining ventricular tissue. The end points of the cables which represented the bundle branches were chosen according to the measurements from Durrer *et al.* [174] (i.e. in the left ventricle, three cables excited the midseptum, the high anterior paraseptal and low anterior papillary muscle regions, and the posterior papillary muscle region whereas a single cable excited the following parts of the right ventricle: the lower septum, apex, and free wall in the region of the anterior papillary muscle).

- Berenfeld *et al.* [184] used a sheet network to model the ventricular conduction system. Short fibers modeled the connections between the sheet and the remaining myocardium. No detailed anatomical description was provided.

- Werner *et al.* [185] used a semi-automatic procedure that is explained in detail in [1, 139, 185] and in section 4.1. It was used as basis for the model of the excitation conduction system that was used in this thesis.

- Ijiri *et al.* [186] proposed the use of an extended, so-called L-system to model the mesh structure of the Purkinje fiber network semi-automatically. The L-system is a formal grammar which was initially introduced by Lindenmayer to describe the development of plants [187] (since then it has been used in a variety of applications to describe complex branching structures; for details see [186]). According to Ijiri *et al.* [186] the conduction system is created by the following steps: at first the endocardial surface has to be chosen, then the user specifies initial points from which the Purkinje fibers will grow. The growth of the Purkinje system is constraint by two extensions of the L-system: the first enforces a uniform Purkinje fiber distribution and the second allows the construction of closed mesh structures (loops). As the growth of the Purkinje fiber network was rapid, the effects of different parameters could be examined by trial and error.

- Zimmerman and Romero *et al.* [188, 189] basically adopted the L-system approach from Ijiri *et al.* [186]. The main differences compared to the approach from Ijiri *et al.* [186] is that they added non-deterministic decisions, which means that the PMJs are not necessarily distributed in a uniform pattern (although no justification was given for this model feature). In addition to that, the proposed method was not able to model loops although this was an important feature of the algorithm from Ijiri *et al.* [186]. The activation of each PMJ was derived by calculating its distance to the AV node and assuming a constant propagation speed of 3 m/s. The resulting excitation sequence did not agree with the measurements from Durrer *et al.* [174] (although this was claimed by the authors).

  In a consecutive project, the same group used splines to connect the PMJs, which have been placed in a pattern that aimed at reproducing the activation sequence of Durrer *et al.* [174] and others (no details were given on the method used to place the nodes) [190]. The electrophysiological modeling of the Purkinje system was adopted from Vigmond *et al.* [180].

  It is interesting to note that the same group did not use their recently developed model in a study which aimed at personalizing the excitation conduction system [191]. In this case they rather resorted to a much simpler approach: they just varied the number of PMJs and their activation time in different regions of the ventricles based on a numerical optimization scheme. The results were not convincing: while the total activation time was closer to the ground truth with a personalized Purkinje system than without any excitation conduction system, the error in the left ventricle was even increased if the excitation conduction system was added to the model.

- Bordas *et al.* [192] used high resolution MR images (of 4 rat and 1 rabbit ventricles) in combination with a semi-automatic segmentation procedure that was able to extract the free-running Purkinje system (i.e. the part of the Purkinje system that runs through the chambers of the ventricles). The reconstructed Purkinje systems exhibited large inter-individual variability in both location and density of the fibers. As the terminal Purkinje network was not visible in the MR images (which is also a current limitation of the approach), the PMJs were placed at the distal ends of the free-running Purkinje system in order to evaluate the effects of these specialized fibers on the sequence of activation in a simulation study.

  The extraction of the Purkinje fiber system from these high resolution MR images was also recently covered in [193]. In this case, the use of *orientation distribution functions* (ODF) was proposed to replace the time-consuming manual extraction.

Some of the models of the excitation conduction system allowed retrograd activation of the Purkinje system (i.e. the depolarization wave travels from the myocardium into the Purkinje network). Although retrograd activation is not important for a physiological activation in sinus rhythm, it should be considered e.g. in models of bundle branch block where the activation front can enter the distal ends of the blocked bundle branch (see [181]). Examples for models that allow such retrograd activation can be found in [194, 195, 196, 181, 180, 190, 184].

Finally, there are a number of electrophysiological models that describe the unique electrophysiological properties of the Purkinje cells. The first model of this kind has been presented by Noble *et al.* [197] and successive models have continuously evolved in both complexity and realism [198, 199, 181, 200].

3.1.5 Modeling Beta-Adrenergic Regulation

As explained in section 2.2.4, beta-adrenergic signaling plays a crucial role in the chronotropic, dromotropic, lusitropic and inotropic regulation of the heart. These effects can be considered in electrophysiological models in a straightforward way: i.e. by simply adapting the conductivity or availability of affected channels accordingly. This strategy was pursued e.g. in [201, 202].

Yet such an approach can not consider potential differences in time scales between the individual components of the adrenergic signaling pathway (slow vs. fast dynamics) or between the adrenergic effects and successive regulatory systems (e.g. calcium/calmodulin dependent kinase (CaMK) phosphorylation [203]). In addition to that, the interplay between changes in cycle length (due to chronotropic effects of ISO on the SN) and ISO-induced APD adaption on the level of the working myocardium might play an important role for the genesis of arrhythmic potential in patients suffering from LQTS [109].

In 2002 the Alliance for Cellular Signaling was formed which aimed at answering global questions about signaling networks [204]. This large-scale collaboration focused on the pathways of two cell types: B lymphocytes and cardiac myocytes. Shortly afterwards, Saucerman *et al.* was the first to publish a model of the entire beta-adrenergic signaling pathway from ligand to various effectors [205]. In this model, transmembrane currents and the intracellular calcium handling were based on Luo Rudy's model of guinea pig's ventricular myocytes (adapted for rabbit by Puglisi *et al.* [206]) which was first adapted to rat [205] and later to rabbit electrophysiology [100]. Other model components were modified based on:

- Experimental calcium handling data from Bers [207]
- Formulations for the L-type calcium channel from Jafri *et al.* [208]
- Formulations for the transient outward potassium current from Pandit *et al.* [209]

The initial version of the Saucerman model [205] was subsequently expanded to consider additional targets of beta-adrenergic regulation (i.e. effects on the ryanodine receptor complex (RyR) and troponin I (TnI) as implemented in [210] and regulation of I_{Ks} as added in [100]). Various versions of the Saucerman model were also used by other groups to model chronotropic [211] and inotropic [212] effects of beta-adrenergic activation as well as its role on the LQTS [109].

Recently, the interplay between beta-adrenergic regulation and the calcium/calmodulin dependent kinase (CaMK) has come into the focus of the scientific community [213, 203]. It is generally assumed that an increased heart rate raises intracellular calcium (Ca^{2+}) levels (partly due to beta-adrenergic regulation), which activate CaMK. However, this additional signaling pathway and its dependency on the adrenergic regulation cascade is outside the scope of this thesis. The interested reader can find a detailed overview describing calcium cycling and signaling in [214].

3.1.6 Investigations Concerning the Congenital Long-QT Syndrome

As explained in section 2.2.5, the congenital LQTS is an ion channel mutation whose investigation during the last 30 years has significantly advanced our understanding of cardiac electrophysiology. Besides phenomenological and statistical discoveries, important findings have also been enabled by experimental and in-silico studies.

When it comes to experimental investigation, the work from Antzelevitch and coworkers is undoubtedly the most famous [215, 216, 217, 218, 219, 220, 221, 222, 223]. In their so-called ventricular wedge experiments, they used an arterially perfused cutout from the left ventricle of a dog which allowed the simultaneous recording of action potential waveforms (from endo, M and epicardial cells) and the corresponding transmural ECG. In this setting, they used pharmacological agents to emulate the effects of the three most important types of LQTS (LQT 1: Chromanol 293B; LQT 2: d-Sotalol; LQT 3: ATX-II). In addition to that, other agents were administered and their effect on the shape of the AP morphology and ECG was evaluated (i.e. Isoproterenol was administered to mimic the role of increased sympathetic activity [110] or Mexiletine and Propranolol were infused to analyze the effects of sodium channel blockers or beta-blockers [217]).

Although several important findings (such as the link between TdP and APD dispersion and many others) originate from these wedge experiments, they also have the following significant limitations:

- The pharmacological agents are an oversimplified electrophysiological model of the real LQTS. The degree of channel block is not known (only the dose of the agent) and neither biophysical changes nor potential adaptation processes that might have been induced by the mutation can be considered.
- The wedge is an oversimplified anatomical model of the complete heart. It is not able to consider potential effects of apico-basal electrophysiological heterogeneities (for details see section 6.2.1). Thus, the transferability of the results from the wedge experiments (e.g. the transmural ECG) is questionable at best.
- Finally, the ECG waveforms extracted from the wedge do not match with the classical waveforms (see Fig. 2.8) that are expected for patients with LQT 1-3 (although the authors usually claim that this is the case).

On the other hand, the published in-silico studies on the LQTS can be separated based on whether they investigated arrhythmogenic mechanism of LQTS in single cells (e.g. [224, 225]) or if they considered action potential propagation and calculated the corresponding ECGs [226, 133, 227, 228]. All computational studies that tried to reproduce the large T-Waves seen in LQT 1 patients failed to do so. In all cases, the reduction of the heterogeneous current I_{Ks} homogenized the repolarization thereby reducing the T-Wave amplitude [226, 133, 227]. This led Potse *et al.* to the unexpected conclusion that I_{Ks} can therefore not be responsible for the APD dispersion and T-Wave concordance in the human ventricle [226]. Although this seems to be far-fetched as other mechanisms might be involved, it is at the same time difficult to rebut his argumentation.

Recently, several computational studies were published that use the model of beta-adrenergic signaling from Saucerman *et al.* (see section 3.1.5) in order to consider the effects of sympathetic influence during the modeling of LQTS. In two cases, this model is used to describe the effects of a previously described AKAP binding domain mutation (see section 2.2.5) [100, 109]. Finally, the Saucerman model has also been used to re-evaluate the effects of beta-adrenergic activation and block in LQT 3 [229].

3.2 Anatomical Modeling and the Forward Problem of Electro-cardiography

3.2.1 Resolution Effects of Anatomical Models

During the last decade, quantitative cardiac modeling has made significant progress with respect to model precision, possible application areas and its clinical significance. These advances are partly due to the increasing computational resources which nowadays allow to model large patches of cardiac tissue or even a whole ventricle using reaction-diffusion models (see section 3.1.3). In contrast to cellular-automaton approaches, which were used in the past [185], reaction-diffusion models consider the coupling of the myocytes through gap junctions which is especially important for the simulation of a realistic repolarization sequence [162].

On the downside, it is known, that the results of reaction-diffusion models depend on the resolution of the underlying anatomical model. In case of low spatial resolution, the conduction velocity of an electrical wavefront is reduced [166]. It is not uncommon to compensate for this effect by adapting the intracellular conductivities accordingly (see e.g. [166, 15, 18]). In case of lower spatial resolutions, the intracellular conductivities must be raised to achieve realistic conduction velocities. Yet even in this case, it is not possible to compensate for the effects that low spatial resolution has on the wavefront curvature. For anisotropic simulations it has been shown that not only the speed but also the shape of the excitation pattern depends on the spatial resolution of the anatomical model [230]. The realistic description of wavefront curvature becomes especially important if waves emerge from narrow tissue structures or propagate around a zone of conduction block during spiral wave rotation [231].

3.2.2 Modeling Ventricular Fiber Orientation

3.2.2.1 General Considerations

As explained in section 2.1.2 the consideration of fiber orientation anisotropy is important for realistic models of the ventricular activation sequence. So far, it has not been possible to create individual or patient-specific models of ventricular fiber orientation. Therefore the fiber orientation information that was measured ex-vivo is usually transferred to the anatomical model under investigations by rule-based [166, 169] or atlas-based techniques [232, 233]. The ex-vivo fiber orientation is usually assessed by histological sectioning or DTMRI as explained in section 2.1.2.

Although DTMRI data has been analyzed at the Institute of Biomedical Engineering (Karlsruhe Institute of Technology) [37, 43, 234, 235] and techniques have

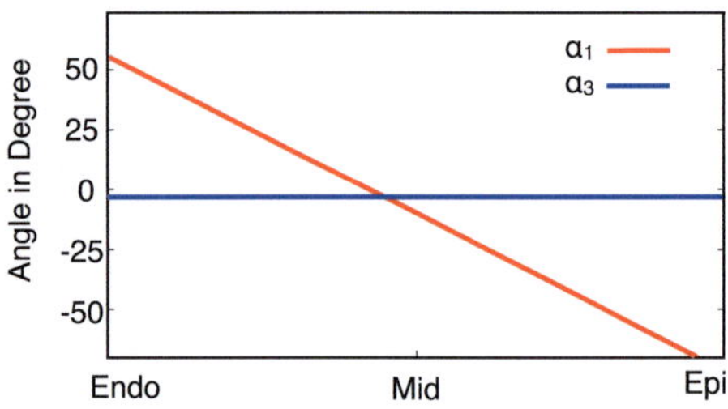

Fig. 3.3. Fiber orientation rules based on the measurements from Streeter *et al.* [35]. These rules were used as input for the algorithm developed by Weiss [38], which we used to create fiber the orientation information for this study.

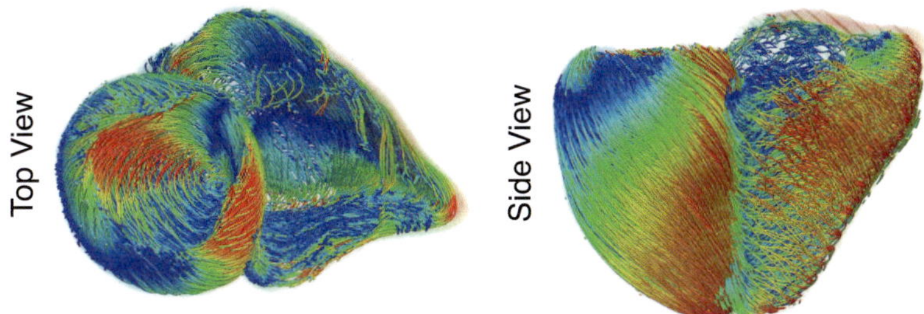

Fig. 3.4. Exemplary visualization of the fiber orientation information that was generated for the ventricles of a healthy 27-year old proband (patient/proband-ID 3; see section 7.1.1). In this case, the x-component of the fiber orientation vector (Cartesian representation) is color coded based on an HSV colormap (red: negative x-component, blue: positive x-component). The streamline visualization was provided by Martin Krueger (IBT).

been developed that allow to transfer this DTMRI-based fiber orientation to arbitrary anatomical models [37, 236] the fiber orientation information used in this study was created with a rule-based approach for reasons of simplicity.

In the future, it will probably become possible to use DTMRI to image the fiber orientation in individual patients in-vivo. Recently, advances that support this assumption have been made [237].

3.2.2.2 Rule-Based Method to Create Fiber Orientation Information

As explained before, we used a rule-based approach developed by Weiss [38] to implement fiber orientation information into our in-silico models of the ventricles. The rules were extracted based on the measurements from Streeter *et al.* [35]. In this case, the helix angle α_1 varied almost linearly from $-75.3°$ at the epicardium to $55.5°$ at the endocardium. In addition to that, a small apico-basal variation in the transverse angle α_3 was reported ($\alpha_{3,Apex} = -3°$ vs. $\alpha_{3,Base} = +3°$) which was discarded in the rule-based approach. Sheets (i.e. the laminar architecture which was described in section 2.1.2) were also not considered. The final course of the helix and transverse angle that was used as input data for the fiber orientation implementation can be seen in Fig. 3.3. No distinctions were made between left and right ventricle (i.e. the same rule was used throughout both ventricles). Details on the implementation of the rule-based approach can be found in [38].

An exemplary fiber orientation dataset that results from the use of the rule-based approach is visualized in Fig. 3.4. In this case, fiber orientation information was generated for the ventricles of a healthy 27-year old proband that was used for different investigations throughout this work (patient/proband-ID 3; see section 7.1.1).

3.2.3 The Forward Problem of Electrocardiography

The forward problem of electrocardiography describes the connection between the electrical sources within the myocardium and the corresponding potential distribution on the body surface. It is solved in various studies throughout this thesis e.g. to predict the effects of heterogeneous ion channel distributions on the T-Wave (see section 4.2) or to evaluate the effects of changes in the excitation conduction system model on the QRS complex (see section 4.1).

3.2.3.1 Problem Formulation and Implementation

In theory it is possible to use the bidomain model to calculate action potential propagation in the tissue (see section 3.1.3). In this case we would directly receive the extracellular potentials in the whole torso model. Furthermore, such an approach would consider the conductivities of the tissues that are surrounding the heart. This can be important for certain applications as changes in the conductivities of these tissue are known to change the distribution of epicardial potentials [238]. However this so called fully-coupled approach also has disadvantages:

- If the bidomain model is used to calculate action potential propagation in the tissue it is important to use small time increments (e.g. $10\text{-}20\mu s$) to obtain realistic simulation results. However, in case of the ECG (extracellular potentials) it is often sufficient to have a relatively coarse temporal resolution (e.g. 1 ms). This means, that a fully-coupled approach is computationally very expensive as the extracellular potentials are calculated more often than necessary (and especially this part of the bidomain model which necessitates to solve a large linear system of equations is computationally very expensive).

- In addition to that, the action potential propagation is simulated on a structured voxel grid at the Institute of Biomedical Engineering (Karlsruhe Institute of Technology). As structured grids do not allow regions with different spatial discretization within an anatomical dataset, the whole torso model would have to be discretized with a resolution that is sufficient for realistic action potential propagation simulations (see section 7.2). In case of a resolution of 0.4 mm^3 this would lead to $\approx 800 \cdot 10^6$ cubic elements and in case of 0.2 mm^3 it would

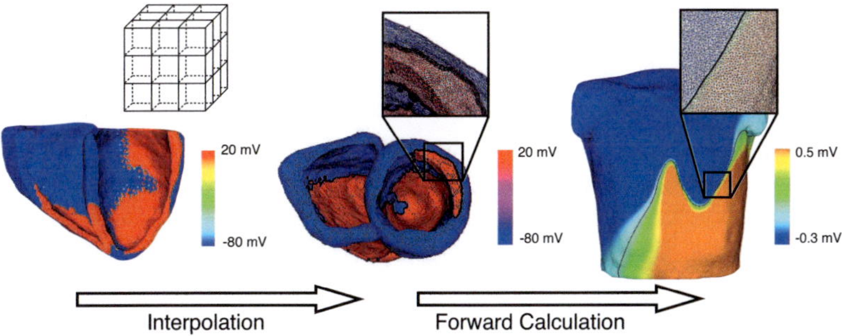

Fig. 3.5. The action potential propagation is simulated on a structured grid using the monodomain model (left side). The resulting transmembrane voltages are then interpolated onto a tetrahedron model of the ventricles (middle) and a part of the bidomain equations is used to calculate the body surface potentials (right side).

even lead to $\approx 6.4 \cdot 10^9$ elements. Simulations on anatomical models of this size are currently not feasible.

To overcome these problems, we use a technique which is often referred to as the two-step approach for solving the forward problem of electrocardiography [238]. In this case, the source terms (transmembrane voltages) are calculated in the uncoupled heart using the monodomain equation. Here at the Institute of Biomedical Engineering (Karlsruhe Institute of Technology) we then interpolate these transmembrane voltages onto an unstructured tetrahedron mesh for reasons of computational efficiency (see Fig. 3.5). Then the bidomain model is used for the forward calculation as it connects the transmembrane voltage V_m and the extracellular potential Φ_e:

$$\nabla \cdot (\sigma_i \nabla V_m) = -\nabla \cdot ((\sigma_i + \sigma_e) \nabla \Phi_e) \tag{3.17}$$

Here, σ_i and σ_e are the volume-averaged conductivity tensors of the intra- and extracellular domain. The finite element method in conjunction with Dirichlet (reference potential) and Neumann (thorax-air boundary) boundary conditions was used to transform (3.17) into a system of linear equations, which was then solved by applying Cholesky decomposition and a conjugate gradient method. Details on the implementation and solution of the forward problem can be found in [239, 146].

3.2.3.2 Influences of Tissue Conductivities

It is well known, that the solution of the forward problem does not only depend on the position and amplitudes of the cardiac sources but also on the size and location of the internal organs and structures which are often referred to as inhomogeneities [240]. Apart from this geometrical information (size and location) the inhomogeneities are also known to vary with respect to their conductivities and degree of anisotropy [241, 242]. This makes it difficult to predict their influence on the ECG intuitively.

In recent years, efforts have been made to characterize the influence of inhomogeneities on the computed body surface potential maps (BSPMs). Most of the associated simulation studies were based on dipole sources [240] or measured epicardial potentials [243]. However, all studies that were published so far neglect to address the lack of consensus in the literature concerning measured tissue conductivity values [244]. They usually simply choose an arbitrary set of conductivities from the large body of published values. Yet the conductivities of all major inhomogeneities in the thorax differ by a factor of 2.3 to 16.5 between different measurement studies. An example for this conductivity variation is the data published on kidney tissue, which has been reported to have a conductivity of 0.0544 S/m in [245] compared to 0.9 S/m in [246].

There are several reasons for these large conductivity deviations between different studies:

- It is technologically challenging to measure conductivities in the low frequency range which is required to satisfy the requirements of the quasistatic electrocardiographic calculations.
- Measurements are often conducted using different experimental techniques [247]. Other deviations result from measurements on different species or even sample variations within the same species.
- Tissue conductivities change ex-vivo after the sample has been excised [248, 249]. Reasons for this are temperature changes, biological degradation, a changing water content or the onset of ischemic effects [250].
- Pathological conditions can cause changes in tissue or fluid conductivity [251]. Examples for this are cystic fibrosis or pulmonary emphysema, which reduces lung conductivity whereas Pompe's disease leads to low skeletal muscle conductivity. In addition to that, even blood conductivity is variable as it depends on the hematocrit.

So there is an obvious need to evaluate these effects with respect to simulated ECGs as large conductivity uncertainties could introduce significant modeling errors even in case of organs with lower importance. Another aspect that is directly related to the influence of conductivity variations in different inhomogeneities is the required complexity of a torso model. When it comes to the creation of patient-specific anatomical models it is important to know which structures have to be included to achieve realistic results. All remaining inhomogeneities can potentially be removed which would speed up the labor intensive and time consuming process of creating computational models for the solution of the forward problem.

3.2.3.3 Rule-Based Modeling of Skeletal Muscle Fiber Orientation

It has been shown in several different studies that skeletal muscle fiber orientation is important for the realistic solution of the forward problem of electrocardiography [15, 243]. However, if the forward problem should be solved for patient-specific anatomies there is usually no information available concerning the muscle fiber arrangement in the heart or skeletal muscles. The main reason for this is that Diffusion Tensor MRI, which is currently the only technique able to deliver in-vivo data on fiber orientation, is time-consuming and susceptible to motion artifacts. In case of cardiac anisotropy, the missing fiber orientation information is often modeled using rule-based approaches (see section 3.2.2 and [26, 166]). However, regarding the modeling of skeletal muscle fiber direction only few models are available which will be presented in the following:

- The model from Bradley *et al.* is based on an anatomical dummy [240]. Skeletal muscle fiber orientation was extracted from this dummy using a magnetic tracking device. The extracted orientation information was then matched onto the torso model that was used for the simulations by a coordinate transformation which relied on anatomical landmarks. A disadvantage of this model is, that it depends on the fidelity of the anatomical dummy which was never validated. Another limitation is associated with the measurement positions: Fibers could only be tracked on the outer or inner surface of the skeletal muscle layer. Fibers between these surfaces were assumed to change their direction in a linear fashion.

- Johnson *et al.* [252] initially proposed a simplified, rule-based method to determine the skeletal muscle fiber orientation which was subsequently adopted by Klepfer *et al.* [243]. In this case, the torso was considered to be a cylinder around which the muscle fibers wrap. In order to implement this distribution, the torso was divided into twelve segments (from a cross-section perspective) and the fiber direction was assumed to be perpendicular to the bisector of each segment (see [252, 243] and section 5.3). A validation of the associated fiber orientation is lacking here as well.

- Finally Sachse *et al.* created a model of skeletal muscle fiber orientation based on the highly detailed thin-section photos of the Visible Man dataset (National Library of Medicine, Bethesda, Maryland, USA) [253]. In this case, automatic methods such as texture analysis and a 3D Sobel filter were used to extract an initial fiber orientation. In a post-processing step, these initial orientations were manually checked and if necessary revised by human experts. As this model of skeletal muscle fiber orientation was based on image data it is safe to assume that it contains the highest level of detail and the most realistic description of the skeletal muscle fiber orientation that is available.

3.2.3.4 Effects of Ventricular Deformation on the Morphology of the T-Wave

Today, the forward problem of electrocardiography is usually solved using static models of the heart and torso that neglect ventricular contraction and relaxation as well as other movement (e.g. due to respiration). While ventricular deformation does not impact on simulations of the depolarization sequence (as the contraction occurs after the electrical activation of the tissue) both contraction and relaxation occur during the ST segment and the T-Wave of the ECG. It can be assumed that the associated change in distance and relation between the cardiac source and the ECG electrode position at the body surface is likely to have effects on simulated BSPMs.

This assumption is supported by several in-silico studies that have shown that both the position and orientation of the heart inside the thorax have a major impact on body surface potentials [254, 251]. It is also backed up by experimental studies: In a clinical trial, Feldman *et al.* infused 15 normal subjects with methoxamine and performed Valsalva maneuver to increase and decrease left ventricular dimensions [255]. In these experiments, the T-Wave amplitude depended directly on the size of the ventricular chambers. In contrast to that, no significant effects were seen if the proximity of the left ventricle to the thoracic surface was altered. The changes in T-Wave amplitude were attributed to alterations in endocardial to epicardial surface ratio which were caused by the thinning of the ventricular wall.

Thinner walls led to an increased transmural gradient during ventricular repolarization and an augmentation of the endocardial influence. The result was a higher T-Wave amplitude. In addition to that, it has been shown that a larger blood volume in the ventricles leads to increased body surface potentials [256]. This has become known as the Brody effect [257].

In recent years, electromechanical models have also been used to estimate the changes in the body surface ECG that is associated with ventricular motion. In a study by Xia *et al.*, the ventricular activation was calculated with a simplified cellular automaton whereas the resulting mechanical deformation was modeled based on composite material theory and the finite element method [258]. Subsequently, the same research group constructed a dynamic heart model based on MR image data [259]. However, the results of this study remain questionable as the dynamic heart model was put in a widely used standard torso model to calculate the ECG. If the heart was not aligned correctly in the standard torso model (which can be difficult as both position and orientation have to be correctly estimated), changes due to ventricular deformation could have been over- or underestimated.

Finally, Smith *et al.* investigated the effects of deformation and mechanoelectrical feedback on the ECG in a 2D slice of the human heart and torso [260]. In this case, the deformation was reported to reduce the T-Wave amplitude while the QT time was shortened due to stretch activated channels.

Part II

Methods

4

Electrophysiological Modeling

This chapter introduces the methods that were used for the various *electrophys-iological* studies conducted during the course of this thesis (i.e. the modeling of the specialized excitation conduction system, investigations on the effects of elec-trophysiological heterogeneities on the T-Wave, the role of beta-adrenergic regu-lation on ventricular electrophysiology and investigations on the congenital LQT syndrome). The corresponding results can be found in chapter 6.

4.1 Modeling the Specialized Excitation Conduction System

As explained in section 2.2.3 the specialized excitation conduction system is a major determinant for the ventricular activation sequence and thus has to be in-cluded in any realistic model of ventricular excitation. Although a number of modeling studies have included descriptions of the terminal Purkinje fiber net-work (for details, please see section 3.1.4) most of the proposed models can not be easily transferred to different anatomical datasets (i.e. most of the models were manually constructed and tailored to specific anatomies). In addition to that, the existing semi-automatic procedures, which would allow such a transfer, have so far mostly not been evaluated with respect to their ability to generate realistically shaped QRS complexes.
In this context, the following study aims at creating a realistic endocardial stimu-lation profile that characterizes the location and time instant of ventricular stimu-lation. The underlying semi-automatic approach facilitates an easy transfer of the resulting stimulation profile to different anatomical datasets by adapting certain model parameters. Furthermore, we conducted a sensitivity analysis to evaluate the importance of the various parameters towards the creation of realistic exci-tation sequences. Finally, body surface potential maps are computed and the ex-

tracted Einthoven leads are compared to clinically acquired ECGs (on the same volunteer) to further validate the proposed method. The results of this study can be found in section 6.1.

In addition to the study which is presented in the following, the proposed endocardial stimulation profiles were also used in a paper presented by Kalayciyan *et al.* [3].

4.1.1 Anatomical and Electrophysiological Input Data

The anatomical models used in this study were derived from the MR images of a healthy 27-year-old volunteer (patient/proband-ID 3; see section 7.1.2). Prior to the segmentation of the ventricles (see section 5.1) the ventricular MRI dataset was trilinearly interpolated (from 1x1x1 mm^3) to a cubic voxel side length of 0.4 mm, which resulted in a total number of 352 x 246 x 275 elements. The corresponding torso model had 437 x 500 x 232 cubic elements with a side length of 1 mm. It contained varying tissue conductivities for lungs, liver, spleen, kidneys, skeletal muscle, heart muscle and fat tissue based on the measurements from Gabriel *et al.* [242]. The anisotropic properties of the cardiac muscle fibers were considered both in the thoracic as well as in the ventricular dataset. The underlying fiber orientation was modeled based on rules proposed by Streeter *et al.* [35] (for details, please refer to section 3.2.2).

The ionic model used for describing the dynamic electrophysiological properties of the ventricular tissue was developed by ten Tusscher *et al.* [151]. Ion channel heterogeneities were considered which are important for a realistic sequence of repolarization [67]. In this case, the density of I_{Ks} (in transmural and apico-basal direction) and I_{to} (in transmural direction) was modified. To this end, three distinct tissue layers were considered in transmural direction: endocardium 40%, midmyocardium 40% and epicardium 20%. Furthermore, changes in electrotonic coupling (due to changes in tissue resistivity) through the ventricular walls according to [261] and [262] were modeled by adapting the respective tissue conductivities (based on [262]). In apico-basal direction, it has been reported that apical I_{Ks} is twice as large as at the base of the ventricle [263]. We implemented this gradient in our model, while the I_{Ks}-density in-between apex and base was linearly interpolated. Finally, the density of the homogeneously distributed potassium channel I_{Kr} was increased by 50% in order to account for the faster repolarization that has been

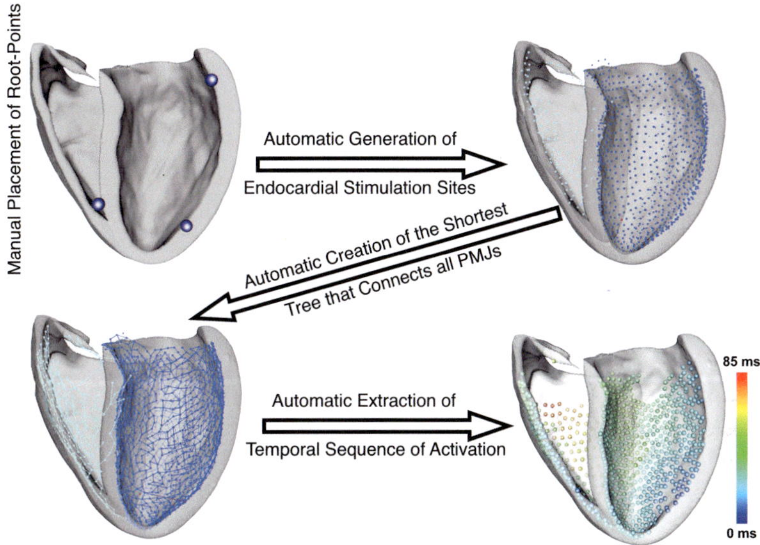

Fig. 4.1. Schematic diagram visualizing the semi-automatic creation of an endocardial stimulation profile. The algorithm is initialized by manually chosen root-points. In a next step the algorithm automatically creates a user-definable number of PMJs at the endocardial surface and connects the nodes to a tree structure. Finally this tree is walked and the algorithm derives a unique stimulation time for each PMJ by calculating the geometrical distance to its root-point and dividing it by a certain conduction velocity.

observed in the clinically acquired recordings. No electrophysiological differences between left and right ventricle were considered.

Before initializing the ventricular activation, all parameter configurations of the electrophysiological model were pre-calculated in an uncoupled environment for a duration of 60 s with a basic cycle length of 0.87 s which matched the proband's heart rate during the clinical recording. Action potential propagation in the cardiac tissue was described with the monodomain reaction-diffusion model (see section 3.1.3). The simulation of ventricular de- and repolarization (400 ms) took about 14h on 14 CPUs (2 GHz Apple Xserve PPCs).

4.1.2 Creation of the Endocardial Stimulation Profile

The semi-automatic procedure that is used to create the endocardial stimulation profile is an extension of previous work from Werner *et al.* [185]. Technical details of the extension can be found in [1]. The proposed method is initialized by manually chosen root-points. Usually their initial location is based on the position of the following anatomical structures: the apical point of the right bundle branch as well as the apical point of both the anterior and the posterior fascicle of the left bundle branch. It can also be advisable to place them near the regions of early endocardial activation that were reported by Durrer *et al.* [174]. In a next step, the algorithm automatically generates a user-definable number of nodes at the endo-cardial surface (see Fig. 4.1). In this case, both the density of the nodes, as well as the minimal distance of the nodes to the atria can be specified. Now, a modified version of Prim's algorithm is used to determine the three sub-spanning trees that contain all endocardial nodes (details on the implementation of Prim's algorithm can be found in [185, 1]). This search is started at the manually placed root-points (see Fig. 4.1). In the following, the algorithm differentiates between two modes:

- In the first case, only end nodes that have no successors are defined as Purkinje fiber endings or Purkinje muscle junctions (PMJs). This option was chosen for this study.
- In the second case, all nodes are defined as Purkinje fiber endings or PMJs (this case is visualized in Fig. 4.1).

At each PMJ, a spherical area of variable size is stimulated to initialize excitation spread. In order to derive the time instant of activation for each PMJ, we calculated the geometrical distance to its corresponding root point and divided this distance by the (user-definable) conduction velocity inside the Purkinje fiber tree structure (see Fig. 4.1). Moreover it is possible to consider potential delays that might occur within the left or right bundle branch by introducing a time-offset for each of the three root-points (e.g. slower conduction velocity in one of the three trees can be modeled by specifying an adequate time-offset for the corresponding root-point). Finally, clinical activation time data or isochrone measurements from the literature can be considered by manually adding additional PMJs. Fig. 4.2 gives an overview over the most important parameters that can be adapted to ensure the creation of a realistic stimulation profile.

In order to evaluate the effect of these different parameters on the QRS complex in the ECG, a standard stimulation profile setup was generated based on which all following stimulation profiles were derived by varying the respective parameter that is under investigation. This standard stimulation profile was created with the following parameters:

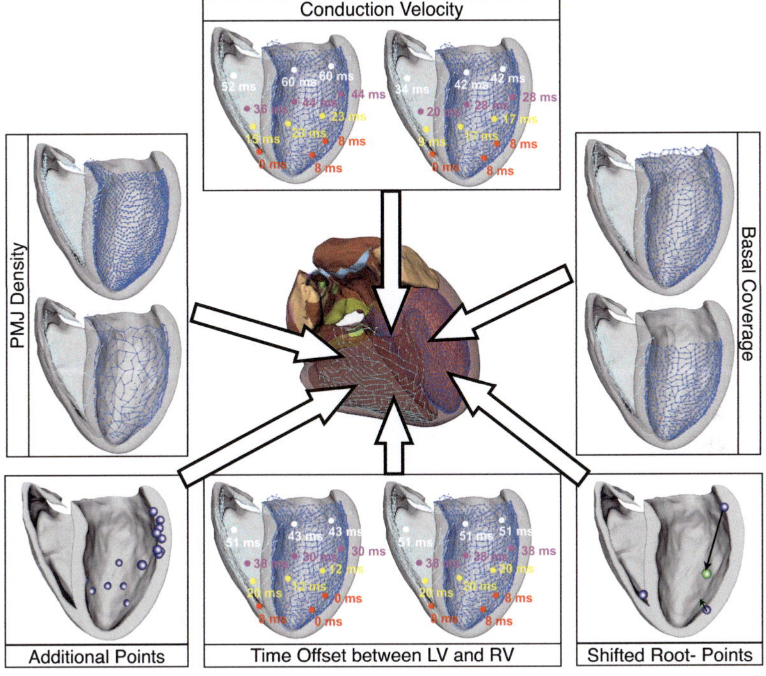

Fig. 4.2. Overview showing a selection of the most important parameters that can be adjusted during the creation of an endocardial stimulation profile.

- The conduction velocity inside the Purkinje tree structure in both left and right ventricle was set to $v = 2000\,\text{mm/s}$
- The minimal distance between the nodes of the Purkinje tree was set to 3 mm (left ventricle) and 5 mm (right ventricle) thus generating 744 PMJs in total.
- The basal part of the ventricles was not covered with Purkinje nodes: left ventricle (15% uncovered), right ventricle (20% uncovered).
- The temporal offset between left and right root-points was chosen such that the left root-points were stimulated 8 ms after the right root-point. The reason for this was the greater distance between the AV node and the left root-points.
- The left posterior fascicle was shifted to the basal paraseptal wall to one site of early activation as recorded by Durrer *et al.* [174].

- A spherical area containing 60-115 voxels was stimulated at each PMJ. The stimulation amplitude was maximal at the center of the sphere and decayed towards its borders according to a Gaussian envelope.

4.1.3 Forward Calculations and Clinical ECG Recording

In order to be able to analyze the effects of a differently parameterized stimulation profile on the QRS complex of the ECG, we had to solve the forward problem of electrocardiography. To this end, a tetrahedron mesh of the voxelized thorax dataset was created on which the simulated transmembrane voltage distributions could be interpolated. Then the bidomain theory was used to derive the corresponding BSPMs as explained in section 3.2.3. The calculation of 400 ms of BSPMs (1 ms time increment) was finished in approximately 2 hours on a single Xserve CPU.

To further evaluate the realism of the simulated ECG signals, we compared them to a clinically acquired multi-channel ECG recording. In this case, the data was recorded with a 64 lead-system from BioSemi (ActiveTwo) at rest. To ensure that the simulated ECGs were extracted from similar measurement locations as used during the clinical recording, the electrode positions were localized with an electromagnetic tracking system from Polhemus (FASTRAK). The resulting position information was then manually matched to the model of the torso and subsequently used to derive the correct electrode positions. The ECG recording was post-processed by removing the baseline wander [264]. Furthermore the signal was denoised by relocating the reference point from the right iliac fossa to Wilson Central Terminal [265]. In order to create a template heart beat which could be used for comparison, we averaged 340 consecutive heart beats on all channels. The resulting template heart beat can be seen together with the simulated signals in section 6.1.

4.2 Electrophysiological Heterogeneities in the Ventricles

As explained in section 2.2.2 ion channel heterogeneities are hold responsible for the dispersion of repolarization (DOR), which determines the shape and polarity of the T-Wave. Yet due to the complex nature of the underlying processes it is difficult to predict the effects of these ion channel heterogeneities on the ECG without the help of quantitative models of ventricular repolarization.

In this study we want to present such a model. Data on heterogeneous ion channel densities and APD distribution were gathered and used to parameterize a model of human ventricular electrophysiology. As APD dispersion in the human ventricles is thought to be mainly determined by a heterogeneous expression of I_{Ks} channels [66] we chose the g_{Ks}-parameter to consider this dispersions in the model. In the following we calculated activation and repolarization of different heterogenous I_{Ks}-distribution models using an anisotropic and high resolution reaction-diffusion model of the ventricles. The resulting transmembrane voltage distributions were used to solve the forward problem of electrocardiography (see section 3.2.3) and derive the corresponding ECGs for the different I_{Ks} models. The realism of the calculated ECGs was evaluated by comparing them to clinically acquired multi-channel ECG data. These recordings were performed on the same volunteer that also provided the anatomical models of the ventricles and torso which were used for the simulated ECGs.

4.2.1 Anatomical Modeling

The anatomical models of the ventricles and torso used in this study were created based on the MRI scans of a healthy 27-year-old proband (see proband with ID 3 in Table 7.1 and Fig. 7.3). The MR imaging sequence and the following segmentation procedure is described in detail in [13] and in section 5.4.2. The diastolic dataset of the ventricles which was used for the simulation of activation and repolarization consisted of 181 million cubic elements (24.5 million elements with cardiac tissue) with a side length of 0.2 mm. Fiber orientation was considered by implementing a distribution based on the measurements from Streeter *et al.* [35]. According to this data, the helix angle α_1 was varied linearly from $55.5°$ at the endocardium to $-75.3°$ at the epicardium while the transverse angle α_3 was kept constant at $-3°$. For more details, please refer to section 3.2.2.

The result of the torso segmentation was a voxelized dataset containing 405.5 million cubic elements with a side lengths of 0.5 mm. For the forward calculations we converted this voxelized dataset into an unstructured tetrahedron model. The meshing process was designed such that we received a higher node density inside the heart than elsewhere. In total, the tetrahedron model of the torso contained 431,449 nodes and 2,435,642 tetrahedrons with an average node distance in the heart of 1.2 mm. The meshing was performed using the Computational Geometry Algorithms Library (CGAL) [266]. Please refer to [13] (section 5.4.2) for a list of all organs that were included in the torso model and a more detailed description of the considered anisotropic properties.

4.2.2 Simulating Ventricular Activation

The electrophysiological properties of the ventricles were characterized using the latest version of the ten Tusscher model for human ventricular myocytes [152]. The model equations were solved with a combination of Rush-Larsen formalism [267] (gating variables) and forward Euler method (all other variables) with a 20 μs time increment. Action potential propagation in the ventricular tissue was described with the monodomain reaction-diffusion (RD) model (see section 3.1.3). The specialized conduction system which initializes ventricular activiation was modeled with a specially adapted endocardial stimulation profile. Details on the construction and parameterization of this profile can be found in [2, 13] and in section 4.1. The stimulation profile parameters were iteratively adapted until the simulated QRS complex was in good agreement with a clinically acquired multi-channel ECG recording. Stimulation profile parameters were:

- Distance between the PMJs was set to 2 mm in the LV and 2.2 mm in the RV.
- One of the LV root-points was shifted to the basal paraseptal wall to one site of early activation as recorded by Durrer *et al.* [174]. No additional, manual stimulation points were added.
- Basal coverage: 5% uncovered in both LV and RV.
- Conduction velocity was set to 2400 mm/s in the left ventricular Purkinje network and to 1900 mm/s in the right ventricular Purkinje network.
- Time offset between LV and RV: the LV was stimulated 8 ms after the RV.

A comparison between the measured and simulated QRS complexes in the Einthoven leads can be seen in Fig. 6.3.

4.2.3 Electrophysiological Heterogeneities

All configurations of heterogeneous I_{Ks} densities that were investigated in this study were based on the literature overview that is displayed in Table 4.1, Table 4.2 and Table 4.3. In the following, we will distinguish between the midmyocardium (Mid) which denotes an anatomical position and the M-cells which have distinct electrophysiological features and do not necessarily only appear in Mid position. The results of the experimental studies were sorted depending on whether they investigated the distribution of the M-cells and commented on transmural (TM) dispersion (see Table 4.1 and Table 4.2) or whether they reported on apico-basal (AB) or interventricular (IV) dispersion (see Table 4.3). Endocardium was denoted Endo and epicardium was referred to as Epi. In total, 19 different heterogeneous I_{Ks} configurations were constructed based on the measurements listed in both tables.

The heterogeneous I_{Ks} distribution was considered during the electrophysiological simulations by storing the maximum channel conductance g_{Ks} in a dataset with identical dimensions and resolution as the anatomical model. This means, that every volume element in the anatomical model had a corresponding g_{Ks} value in this additional dataset. To consider these values during the simulations, the additional dataset was loaded and used to replace g_{Ks} in the electrophysiological model before starting the calculations.

4.2.3.1 Transmural Heterogeneities

All listed studies report a longer Endo APD compared to Epi. This was considered by reducing Epi g_{Ks} ($g_{Ks,Epi} = 0.92 \cdot g_{Ks,Endo}$) according to [268]. Regarding the modeling of the M-cells, the lack of information concerning the exact position, the volume fraction of M-cells within the ventricular wall and the topography of the M-cells was problematic. From Table 4.1 it can be seen, that M-cells have been found at different transmural depths throughout the wall (sub-Endo, Mid layers and sub-Epi) and in different topographical formations (layers, clusters/islands). However, the few studies, that report on the volume fraction of the M-cells, agree that they account for $\approx 30\%$ of the ventricular mass [261, 269]. Based on these reported properties, we assumed a constant thickness of the M-cell layer, which was moved to 3 different locations inside the ventricular wall:

- **TM-60**: this setup contained 60% Endo, 30% M and 10% Epi cells (see Fig. 4.4).
- **TM-40**: this setup contained 40% Endo, 30% M and 30% Epi cells (see Fig. 4.4).
- **TM-20**: this setup contained 20% Endo, 30% M and 50% Epi cells (see Fig. 4.4).

Similarly as proposed above, the modeling community usually integrates the M-cells in layers of varying thicknesses into their in-silico models of the ventricles [166, 159, 161, 164]. However, recently several experimental studies proposed the existence of a more complex M-cell topography [155, 154]. In these studies, the M-cells were found to be clustered in the shape of islands. To investigate the effects of a clustered M-cell arrangement on the ECG, we created 2 additional setups. In both cases, the center of the M-cell islands was positioned close to the Endo (distance: 20.5% of total wall thickness) in both left ventricle (LV) and right ventricle (RV) according to [155]. M-cell island size was reported to be 2.4 mm in depth (transmural direction) and 4.7 mm in width (longitudinal direction) [155]. No information was available on the height of the islands as the shape of the M-cell clusters was investigated on the cut surface of a left ventricular wedge preparation [155]. For reasons of simplicity, we assumed equal width and height and modeled the M-cell islands as ellipsoids. The resulting shape, location and size of the M-cell islands was similar to the experimental reports (compare magnification of Fig. 4.4 **TM-IS-4r** with pictures in [155]). In case of the right ventricle no measurement data on the size of the islands was available. Therefore we simply halved the width, height and depth of the M-cell ellipsoids to account for the thinner RV walls. To additionally investigate the effects of M-cell island packing density we created the following two setups:

- **TM-IS-3r**: here, the center of the M-cell islands were spaced 3*r (r = height = width) apart (see Fig 4.4).
- **TM-IS-4r**: here, the center of the M-cell islands were spaced 4*r (r = height = width) apart (see Fig 4.4).

In all transmural setups, the septum was assumed to have similar properties as the LV free wall (e.g. Epi cells on the side of the septum that faces the RV and Endo cells on the side facing the LV) as can be seen in Fig. 4.4. All parameters of the electrophysiological model for the three cell types were adopted from the original ten Tusscher *et al.* publication [152] (except $g_{Ks,Epi}$ which was reduced to 92% $g_{Ks,Endo}$ as mentioned before).

Table 4.1. Overview: Repolarization Heterogeneities in the Ventricles (Part I)

Species		Measurement Position	M-cells	Remarks
Human	[261]	LV free wall	deep sub-Epi	Endo: 60%, M: 30%, Epi: 10%
Human	[54]	RV free wall	Mid layer	no AP notch visible
Human	[155]	posterior-lateral LV free wall	deep sub-Endo (3.4 mm from Endo border)	M-cells had island topography
Dog	[269]	LV free wall ; septum (near base)	deep sub-Epi (LV free wall); deep sub-Endo (septum)	M-cells comprise 20 - 40% of ventricular mass
Dog	[262]	LV anterior wall	deep sub-Endo	2.8 fold resistivity increase from M to Epi
Dog	[270]	LV free wall; RV free wall	deep sub-Epi to Mid (LV); deep sub-Epi (RV)	
Dog	[271]	septum	midseptal region	transseptal dispersion was twice dispersion found in LV free wall
Dog	[272]	LV anterior wall	deep sub-Endo (1-2 mm from Endo border)	M-cell layer parallel to Endo
Dog	[154]	anterior, anterolateral, posterior LV free wall	midwall region	M-cells found in islands or clusters that vary in extent and location
Dog	[273]	anterobasal LV free wall	found in all loci of transmural slab	no M-cells found in-vivo
Pig	[274]	LV free wall	Mid	some M-cells found in Epi
Guinea pig	[275]	LV free wall	Mid	
Rabbit	[276]	LV free wall	sub-Endo to Mid	mixed populations of M and sub-Endo cells

Table 4.2. Overview: Repolarization Heterogeneities in the Ventricles (Part II)

	Species		Measurement Position	Transmural Dispersion	Remarks
Transmural Dispersion	Human	[55]	anterior LV (midway apex, base)	$APD_{Endo} > APD_{Epi}$	I_{to1} larger in Epi vs. Endo
	Human, Dog	[277]	LV halfway on apico-basal axis	$APD_{Mid} > APD_{Epi}$	
	Dog	[278]	LV	$APD_{Endo} > APD_{Epi}$	
	Dog	[279]	posterior LV	$APD_{Endo} > APD_{Epi}$	I_{Kr} larger in Epi vs. Endo
	Dog	[280]	LV posterior base and lateral wall	$FRP_{Endo} > FRP_{Epi}$	functional refractory period (FRP) $\approx$ APD
	Rabbit	[281]	LV free wall	$APD_{Endo} > APD_{Epi}$	M-cells located in Endo
	Guinea pig	[282]	no precise location given	$APD_{intramural} > APD_{Endo,Epi}$	$APD_{Endo} = APD_{Epi}$
	Guinea pig	[283]	base LV free wall	$APD_{midmyocard} > APD_{Endo} > APD_{Epi}$	no M-cells in LV free wall
	Ferret	[284]	LV	$I_{Kr,Epi} > I_{Kr,other\ layers}$	

Table 4.3. Overview: Repolarization Heterogeneities in the Ventricles (Part III)

	Species		Measurement Position	Apico-Basal Dispersion	Remarks
Apico-Basal Dispersion	Human, Dog	[263]	LV free wall (apex vs. base)	$APD_{Apex} < APD_{Base}$	Dog: I_{Ks}, $I_{to} \approx$ double in apex vs. base; Human: smaller gradient
	Dog	[285]	whole LV (various locations: apex vs. base)	$RT_{posteriorapical} > RT_{anteriorbasal}$	repolarization time (RT) = activation time + activation recovery interval, no RT transmural gradient
	Dog	[286]	left anterior wall (apex vs. base)	$ERP_{Apex} > ERP_{Base}$	ERP $\approx$ APD, no transmural ERP dispersion
	Dog	[269]	LV free wall (apex vs. base)	$APD_{Apex} > APD_{Base}$	
	Dog	[278]	LV Endo, Epi (apex vs. base)	$APD_{Apex} > APD_{Base}$ (Endo and Epi)	
	Dog	[280]	LV Epi	$FRP_{Apex} > FRP_{Base}$	functional refractory period (FRP) $\approx$ APD
	Rabbit	[287]	apex vs. base, no location details	$APD_{Apex} > APD_{Base}$	$I_{Ks} \approx$ tripple in base vs. apex
	Rat	[288]	LV Endo (apex vs. base)	$APD_{Apex} < APD_{Base}$	
	Guinea pig	[282]	apex vs. base, no location details	$APD_{Apex} < APD_{Base}$	longest APD in papillary muscles, shortest APD in septum

	Species		Measurement Position	Interventricular Dispersion	Remarks
IV Dispersion	Dog	[56]	Epi base (central RV free wall vs. central LV free wall)	$APD_{RV} < APD_{LV}$	I_{to} larger in RV vs. LV
	Dog	[289]	Endo (RV vs. LV)	$APD_{RV} < APD_{LV}$	random positions on Endo
	Dog	[57]	M-cells (RV vs. LV)	$APD_{RV} < APD_{LV}$	I_{Ks}, $I_{to} \approx$ double in RV vs. LV
	Rat	[288]	Endo (RV vs. LV)	$APD_{RV} < APD_{LV}$	
	Guinea pig	[290]	RV vs. LV	$APD_{RV} > APD_{LV}$	I_B was measured, APD changes predicted using computer model

4.2.3.2 Apico-Basal Heterogeneities

Concerning the distribution of APD heterogeneity in apico-basal direction there are contradicting reports in the literature (see Table 4.3). Both shorter and longer APD has been found at the apex compared to the base. To account for these inconsistent findings we constructed two setups with opposing apico-basal heterogeneities, that will be described in more detail in the following.

One of the three cell types has to be chosen as basis for the apico-basal setups. In this case we assumed that the ventricles contained purely Endo cells (due to reports that the majority of the LV is made up of Endo cells [261]) and assigned the corresponding values for g_{Ks} (0.36064 nS/pF) and g_{to}. Measurements on dogs found apical I_{Ks} to be twice as large compared to the base [263]. For human samples, a similar but smaller apico-basal gradient was found [263]. Based on this data, we created a model with a 1.5 times larger I_{Ks} density at the apex (vs. base) and another model with a 1.5 times larger I_{Ks} density at the base (vs. apex). All g_{Ks} values in between apex and base were linearly interpolated. Both apico-basal setups can be seen in Fig. 4.4:

- $\mathbf{A^1B^{1.5}}$: Endo g_{Ks} was multiplied with apico-basal scaling factors. The result was a higher basal g_{Ks} density ($g_{Ks,Apex} = 0.36064$ nS/pF,
  $g_{Ks,Base} = 0.54096$ nS/pF).
- $\mathbf{A^{1.5}B^1}$: Endo g_{Ks} was multiplied with apico-basal scaling factors. The result was a higher apical g_{Ks} density ($g_{Ks,Apex} = 0.54096$ nS/pF,
  $g_{Ks,Base} = 0.36064$ nS/pF).

4.2.3.3 Interventricular Heterogeneities

In case of the distribution of interventricular heterogeneities, all but one listed studies agree that RV APD is shorter compared to the APD in the LV (see Table 4.3). Such a relationship was reported for Endo [288, 289], M [57] and Epi cells [56]. According to Volders *et al.*, I_{Ks} is approximately twice as large in the RV. Based on this report, we adapted the previously introduced setups of transmural heterogeneity (layered M-cell topography) by doubling RV g_{Ks}:

- **TM-60*IV**: the transmural setup containing 60% Endo, 30% M and 10% Epi cells was adapted by doubling RV g_{Ks} (see Fig. 4.4).
- **TM-40*IV**: the transmural setup containing 40% Endo, 30% M and 30% Epi cells was adapted by doubling RV g_{Ks} (see Fig. 4.4).

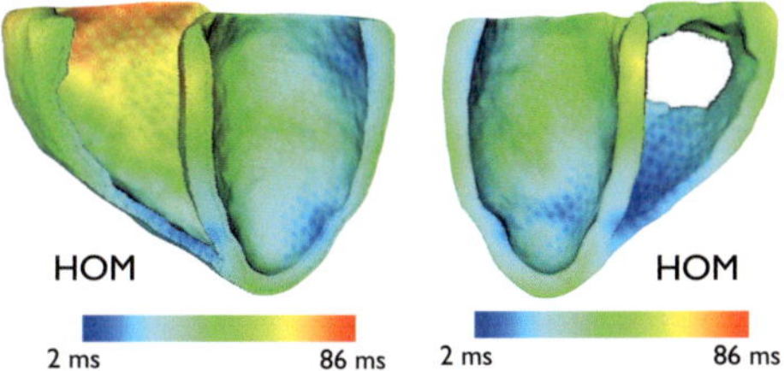

Fig. 4.3. Activation time map of the setup that contained purely Endo cells (**HOM**). Activation was finished after 86 ms. This activation time map was used to construct both setups with a linear relationship between AT and APD (**AT-APD-0.5**, **AT-APD-1.4**). To this end, the AT map was normalized and the resulting scaling factors were multiplied with the g_{Ks} of Endo cells. This means, that areas that were activated later had a higher g_{Ks}-scaling factor which in turn led to a shorter APD_{90}.

- **TM-20*IV**: the transmural setup containing 20% Endo, 30% M and 50% Epi cells was adapted by doubling RV g_{Ks} (see Fig. 4.4).

4.2.3.4 Combination of Apico-Basal and Transmural Heterogeneities

In this section we describe the creation of setups that contained both transmural and apico-basal g_{Ks} gradients. The following 6 setups are a combination of the 3 setups of transmural heterogeneity (layered M-cell topography) and the 2 setups that contain apico-basal heterogeneities:

- **TM-60*A^{1}B$^{1.5}$**: the transmural setup **TM-60** was multiplied with apico-basal scaling factors (apical scaling factor: 1 / basal scaling factor 1.5 / factor in-between: interpolated). The results can be seen in Fig. 4.4.
- **TM-40*A^{1}B$^{1.5}$**: the transmural setup **TM-40** was multiplied with apico-basal scaling factors (apical scaling factor: 1 / basal scaling factor 1.5 / factor in-between: interpolated). The results can be seen in Fig. 4.4.
- **TM-20*A^{1}B$^{1.5}$**: the transmural setup **TM-20** was multiplied with apico-basal scaling factors (apical scaling factor: 1 / basal scaling factor 1.5 / factor in-between: interpolated). The results can be seen in Fig. 4.4.
- **TM-60*A$^{1.5}$B^1**: the transmural setup **TM-60** was multiplied with apico-basal scaling factors (apical scaling factor: 1.5 / basal scaling factor 1 / factor in-between: interpolated). The results can be seen in Fig. 4.4.
- **TM-40*A$^{1.5}$B^1**: the transmural setup **TM-40** was multiplied with apico-basal scaling factors (apical scaling factor: 1.5 / basal scaling factor 1 / factor in-between: interpolated). The results can be seen in Fig. 4.4.

- **TM-20*A$^{1.5}$B^1**: the transmural setup **TM-20** was multiplied with apico-basal scaling factors (apical scaling factor: 1.5 / basal scaling factor 1 / factor in-between: interpolated). The results can be seen in Fig. 4.4.

4.2.3.5 Other Electrophysiological Configurations

Finally, we wanted to consider the studies that found an inverse linear relationship between activation time (AT) and APD [280, 291, 292, 293, 294, 295]. One study even hypothesized that it might be an intrinsic property of the myocardium that enables it to adapt its APD to the activation sequence [295]. Some of these studies performed a linear regression analysis and found an inverse relationship between AT and APD. This means that progressively later activation was associated with progressively shorter APDs. Although all studies agreed on this inverse relationship between AT and APD there are contradictions with respect to the slope of the linear fit. The majority of the studies reported slopes steeper than negative unity (-1.3 [291], -1.32 [293], -1.44 [295]) while one study reported smaller slopes between -0.10 and -0.93 [292]. The steepness of the slope is important as it is directly related to the repolarization sequence. In case of small slopes (< -1) the shortening of the APD is not enough to compensate for the delay in AT. Therefore the sequence of repolarization is similar (same direction) to the sequence of activation. In contrast to that, the repolarization sequence is reversed (opposite direction than the depolarization) if the slope is larger than negative unity as the shortening of the APD is in that case larger than the delay in AT. In order to evaluate the effects of both slopes on the shape of the T-Wave, we created two heterogeneous g_{Ks} distributions that were associated with slopes < -1 and > -1, respectively. Similar as done for the construction of the apico-basal setups, we again assumed that the ventricles consisted out of Endo cells and chose the parameters of the electrophysiological model accordingly. The heterogeneous g_{Ks} distributions for both setups were created based on an AT map of a homogeneous setup that contained purely Endo cells (see Fig. 4.3):

- **AT-APD-0.5**: here, the maximum of the AT map from Fig. 4.3 was normalized to 1.75 (the minimum was normalized to 1). The resulting scaling factors were multiplied with the g_{Ks} of Endo cells ($g_{Ks,Endo} = 0.36064$ nS/pF).
- **AT-APD-1.4**: here, the maximum of the AT map from Fig. 4.3 was normalized to 2 (the minimum was normalized to 0.25). The resulting scaling factors were multiplied with a g_{Ks} of 0.36064 nS/pF.

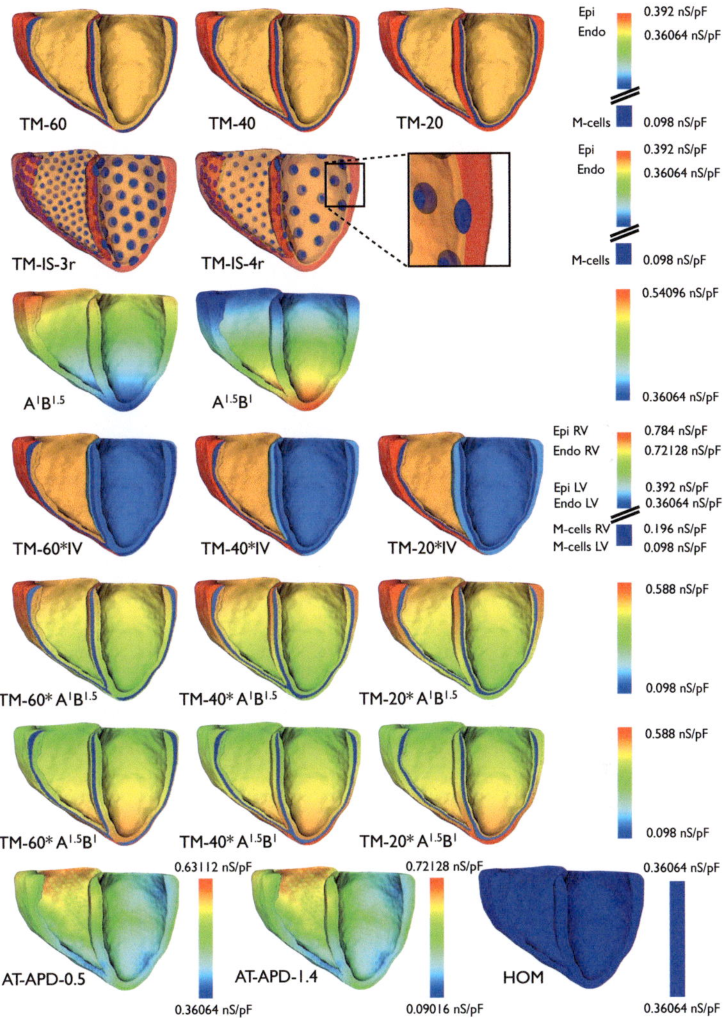

Fig. 4.4. Distribution of the heterogeneous g_{Ks} parameter. In the first and second row g_{Ks} is displayed for the different setups of transmural heterogeneity. Row three shows the apico-basal configurations whereas row four visualizes g_{Ks} for the interventricular setups. Rows five and six show g_{Ks} for the mixed transmural and apico-basal setups while the last row displays the distribution of the remaining electrophysiological configurations.

Due to this approach, areas that were activated later had a higher g_{Ks}-scaling factor. This led in turn to a shorter APD_{90}.

The last setup which was evaluated in this study was a configuration that had a homogeneous g_{Ks} density:

- **HOM**: this setup contained only Endo cells.

4.2.4 Forward Calculation of the ECG

For each heterogeneous electrophysiological configuration that was introduced in section 4.2.3 a simulation of ventricular activation and repolarization was conducted. The resulting transmembrane voltages were then interpolated onto the unstructured tetrahedron mesh of the ventricles that was created as described in section 4.2.1. Subsequently, the bidomain model was used to calculate the extracellular potentials in the whole torso. Tissue conductivities that were used during the forward calculations were based on the measurement data from Gabriel et al. at 10 Hz [245]. The forward calculation procedure is described in more detail in [15] and in section 3.2.3.

To ensure comparability of the simulated and measured ECGs, we extracted the simulated body surface potentials at the same positions that were used during the multi-channel ECG recording. A visualization of these electrode positions can be found in Fig. 5.10. To be able to transfer the electrode position from the clinical measurements to the tetrahedron model of the torso, we localized the electrodes with an electromagnetic tracking system (FASTRAK, Polhemus, Burlington VT, USA). This measured position information was then manually matched to the tetrahedron mesh of the torso.

4.2.5 Multi-channel ECG Recording

The multi-channel ECG data was recorded with a 64 lead-system (ActiveTwo, BioSemi, Amsterdam, Netherlands) at ≈ 70 beats per minute. The raw ECG data was post-processed by removing the baseline wander [264] and increasing signal to noise ratio [265]. Finally, a template heartbeat was created that could be used to compare the simulated ECGs with the measurement. To this end, 116 consecutive heart beats were averaged on every channel which was equivalent to 100 s of ECG recording time.

4.3 Beta-Adrenergic Regulation of Ventricular Electrophysiology

4.3.1 Overview Over the Intracellular Signaling Pathway

In the course of this thesis, we integrated main components of the intracellular adrenergic signaling pathway from Saucerman *et al.* [100] into the latest revision of the ten Tusscher model for human ventricular myocytes [152]. Essential parts of the regulatory pathway will be explained in the following. A more detailed explanation of all biophysical transitions and the underlying equations is provided in [5].

The adrenergic signaling is activated by Norepinephrine (NE) or Isoproterenol (ISO) binding to β1-receptors (see Fig. 4.5: green structure). The receptor-ligand complexes are coupled to G_s proteins, subsequently inducing an adenylyl cyclase (AC) stimulation. AC promotes the synthesis of the second messenger cyclic AMP (see Fig. 4.5: magenta structures) which in turn activates protein kinase A (PKA) (see Fig. 4.5: blue structures). The catalytic subunits of PKA now phosphorylate a number of different target proteins (see Fig. 4.5: red structures) which has direct (e.g. I_{CaL}, I_{Ks}) or indirect (e.g. SERCA, I_{NaK}) effects on the conformation of the channel proteins. In this context, the term phosphorylation describes the most common reversible modification of a protein thereby regulating its activity. The regulation works as follows: the protein kinase can add a phosphate group whereas the protein phosphatase removes it again.

In the latest version of the intracellular signaling cascade from Saucerman *et al.* [100], adrenergic regulation of five target proteins was considered: I_{CaL}, RyR, SERCA, TnI and I_{Ks}. When we transferred the adrenergic signaling pathway to the ten Tusscher model, we did not consider the adrenergic effects on RyR and TnI (see Fig. 4.5: canceled connection between PKA and RyR as well as between PKA and TnI). This was done for two reasons: On the one hand, adrenergic effects on these proteins are controversially discussed (especially with respect to RyR [296, 297, 298]) and on the other hand, extensive adaptations would have been necessary as the corresponding equations differed significantly between the Saucerman and the ten Tusscher model. However, after performing some initial experiments, we added adrenergic regulation of I_{NaK} later on. This became necessary to avoid excessive accumulation of intracellular calcium (for a detailed explanation see section 4.3.2: I_{NaK} and section 6.3.1).

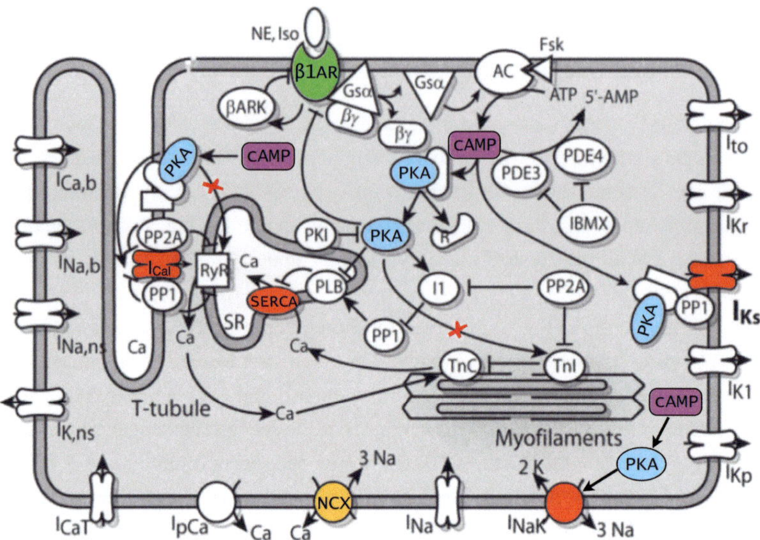

Fig. 4.5. Schematic of the beta-adrenergic signaling pathway which was included into the second revision of the ten Tusscher model for human ventricular myocytes. NE or ISO binds to β1-receptors (green) which eventually raises cAMP levels (magenta). Subsequently, cAMP activates PKA (blue) which phosphorylates target proteins (red) thereby directly (e.g. I_{CaL}, I_{Ks}) or indirectly (e.g. SERCA, I_{NaK}) affecting the channel conductivity or availability. Furthermore, indirect effects on NCX were triggered (orange). This was due to the enhanced removal of sodium from the intracellular space (through the ISO-induced increase in I_{NaK}). A lowered cytosolic sodium concentration enhanced the function of NCX which could then remove more calcium from the intracellular space. Figure modified based on [100].

4.3.2 Target Proteins of Beta-Adrenergic Regulation

In the following, the effects of the beta-adrenergic regulation on the target proteins will be explained and the respective equations will be provided. Generally speaking, the adrenergic influence was modeled by multiplying the channel conductivity with a factor or adding a summand that caused a leftward shift in the current-voltage relationship. If no adrenergic effects were present, the factor was set to 1 and the summand to 0. In this case, the model behaved like the original ten Tusscher model which was used as basis.

The global effects of ISO can be summarized as follows:

- The phosphorylation of I_{CaL} increases the inflow of calcium into the cell. A larger I_{CaL} triggers a higher calcium release current from the SR. The resulting increase in cytosolic calcium concentration leads to an increase in contractility.

- The phosphorylation of the SERCA (I_{up}) increases the speed with which the calcium is pumped back into the SR after the contraction has been initiated. This leads to an increase in lusitropy.
- The phosphorylation of I_{Ks} increases the sodium conductivity thereby shortening the APD.
- The phosphorylation of I_{NaK} indirectly stimulates the sodium / calcium exchanger NCX. The increase in NCX ensures that excessive calcium which might have accumulated in the cytosol (due to the ISO-induced increase of I_{CaL}) is removed.

L-type Calcium Current (I_{CaL})

Unfortunately, the calcium handling was implemented differently in the Saucerman compared to the ten Tusscher model. Therefore several adaptations were necessary to allow a transfer of the adrenergic signaling pathway (for details please refer to [5]). The following equation was used by ten Tusscher to describe I_{CaL}:

$$
\begin{aligned}
I_{CaL} = & g_{CaL} \cdot f_{po} \cdot f_{avail} \cdot d \cdot f \cdot f_{CaSS} \\
& \cdot f_2 \cdot 4 \frac{(V-15)F^2}{RT} \cdot \frac{0.25 \cdot Ca_{SS} \cdot e^{2(V-15)\frac{F}{RT}} - Ca_o}{e^{2(V-15)\cdot \frac{F}{RT}} - 1}
\end{aligned}
\tag{4.1}
$$

In this case, the adrenergic regulation worked through a phosphorylation of the α_{1C}-subunit (LCCa) of the calcium channel protein which increased the channel's open probability and the phosphorylation of the β-subunit (LCCb) which enhanced the channel's availability. From a mathematical point of view, f_{po} and f_{avail} were introduced as interface to the adrenergic signaling cascade. Their values were above 1 if ISO was present and equal to 1 if no ISO was administered. They could be calculated according to:

$$
f_{po} = 0.03 \cdot \frac{frac_{LCCap}}{frac_{LCCapo}} + 0.97
\tag{4.2}
$$

$$
f_{avail} = 0.05 \cdot \frac{frac_{LCCbp}}{frac_{LCCbpo}} + 0.95
\tag{4.3}
$$

Here, $frac_{LCCap}$ and $frac_{LCCbp}$ was the fraction of phosphorylated LCCa and LCCb protein subunits in the presence of ISO. In contrast to that, $frac_{LCCapo}$ and

$frac_{LCCbpo}$ was the fraction of phosphorylated LCCa and LCCb protein subunits in the absence of ISO.

Calcium Uptake Current (SERCA or I_{up})

In case of the SERCA or I_{up} the effects of the beta-adrenergic regulation are indirect: The activity of I_{up} is regulated by phospholamban (PLB) which has an inhibiting effect on the uptake current. The application of ISO now leads to a phosphorylation of PLB thereby reducing its inhibiting influence. Thus the activity of the pump is indirectly promoted.

Mathematically, the adrenergic influence is represented by an increase in the affinity of the pump towards calcium as a result of PLB phosphorylation. To this end, the term that characterizes the cytosolic calcium concentration at which the pump reaches 50% of its maximum conductivity was modified:

$$Km_{up} = Km_{upo} \cdot (0.6 \cdot \frac{frac_{PLB}}{frac_{PLBo}} + 0.4) \tag{4.4}$$

Here, Km_{upo} is the baseline value of the half-saturation constant (from the original ten Tusscher model) whereas $frac_{PLB}$ and $frac_{PLBo}$ is the fraction of non-phosphorylated PLB in the presence and absence of ISO. If ISO is administered, the value of the half-saturation constant Km_{up} is reduced which means that the pump already reaches half of its maximal pump current at lower cytosolic calcium concentrations.

The ISO-dependent Km_{up} was subsequently integrated into the equation for I_{up}:

$$I_{up} = \frac{V_{maxup}}{1 + Km_{up}^2/Ca_i^2} \tag{4.5}$$

Slow Potassium Current (I_{Ks})

In case of I_{Ks}, the activation of beta-adrenergic regulation caused an increase in maximum conductivity and a leftwards shift in the conductivity-voltage relationship.

The equation for I_{Ks} in the ten Tusscher model is given by:

$$I_{Ks} = g_{Ks} \cdot xs^2 \cdot (V - E_{Ks}) \tag{4.6}$$

The beta-adrenergic regulation was incorporated here by adapting g_{Ks} and xs accordingly.

In case of g_{Ks} the constant maximum conductivities of 0.392 nS/pF for endo- and epicardial cells and 0.098 nS/pF for M cells from the original ten Tusscher model were replaced by:

$$g_{Ks} = g_{Kso} \cdot \left(0.2 \cdot \frac{frac_{Iksp}}{frac_{Ikspo}} + 0.8\right) \qquad (4.7)$$

Here, g_{Kso} represents the old ten Tusscher constants whereas $frac_{Iksp}$ and $frac_{Ikspo}$ denote the fraction of the phosphorylated I_{Ks} channels in the presence and absence of ISO, respectively.

The gate xs was given by:

$$xs_\infty = \frac{1}{1 + e^{(xs_\infty(V05)-V)/14}} \qquad (4.8)$$

Here, $xs_\infty(V05)$ replaced the constant value of -5 that was used in the original ten Tusscher publication:

$$xs_\infty(V05) = -5 - \left(1.5 \cdot \frac{frac_{Iksp}}{frac_{Ikspo}} - 1.5\right) \qquad (4.9)$$

According to this, $xs_\infty(V05)$ was equal to -5 in the absence of ISO as the term in brackets is equal to 0. However, in the presence of ISO, the term in brackets increased, leading to a shift of $xs_\infty(V05)$ to more negative values (which can be translated into a leftwards shift of the channel activation curve).

Sodium / Potassium Pump Current (I_{NaK})

Beta-adrenergic regulation of this channel was not implemented in the first version of the model [5, 7]. However, after performing some initial experiments, we soon realized that the additional calcium which was flowing into the cell due to the ISO-induced increase of I_{CaL} can trigger calcium sparks from the SR (see section 6.3.1). A similar observation was made by Despa et al. [299] who studied the effects of adrenergic regulation on I_{NaK} using genetically altered PLM knockout mice. If no ISO-induced increase of I_{NaK} was possible (as this part of the signaling cascade has been genetically removed) the calcium spark frequency increased significantly. This was attributed to the indirect effects that an increase of I_{NaK} had on the sodium / calcium exchanger NCX. An increase of I_{NaK} removed more sodium from the cytosol which in turn increased the sodium concentration gradient that was used by NCX to remove excessive calcium from the cytosol (for more details see section 6.3.1).

ISO effects on I_{NaK} were implemented in a straightforward way: both the maximum conductivity k_{NaKIso} and the half saturation constant of sodium (K_{mNaIso}) were modified. Both parameters were assumed to be controlled by the proportion of activated protein kinase A I (PKACI). The differences between PKAI and PKAII are explained in [5].

The modified equation of I_{NaK} was defined by:

$$I_{NaK} = k_{NaKIso} \cdot \frac{K_o \cdot Na_i}{(K_o + K_{mK}) \cdot (Na_i + K_{mNaIso})}$$
$$\cdot \frac{1}{(1 + 0.1245 \cdot e^{-0.1 \cdot (VF/RT)} + 0.0353 \cdot e^{-(VF/RT)})} \qquad (4.10)$$

In this case, the new ISO-dependent maximum conductivity of the pump was determined by:

$$k_{NaKIso} = k_{NaK} \cdot \left(fac_{kNaK} \cdot \frac{PKACI}{PKACI_{init}} + (1 - fac_{kNaK}) \right) \qquad (4.11)$$

Here, higher levels of activated PKACI increased the conductivity compared to the baseline value (without ISO) of k_{NaK} that was adopted from the original ten Tusscher publication.

Concerning the half saturation constant of sodium (K_{mNaIso}), the corresponding equation was modified such that the constant had a smaller value in case of an increased level of active PKACI. A smaller half saturation constant enabled the pump to start working at lower levels of cytosolic sodium.

$$K_{mNaIso} = K_{mNa} \cdot \left(fac_{K_{mNa}} \cdot \frac{PKACI_{init}}{PKACI} + (1 - fac_{K_{mNa}}) \right) \qquad (4.12)$$

4.3.3 Technical Implementation

In general, the equations of electrophysiological models that contain transmembrane currents as well as a model for the intracellular calcium handling are solved with a combination of Euler integration and Rush Larsen formalism [267]. An overview over different electrophysiological models and possible solution techniques is given in [46]. However, after we included the adrenergic signaling pathway (which had several interfering loops and processes with different time scales) into the ten Tusscher model, we had to use a solution approach that could handle such a system of equations more efficiently. In this case, we chose a solver from the SUNDIALS suite [300] which is able to adapt its step size depending on the degree of change in the solution function (variable-step, multi-step method).

In addition to that, parts of the signaling pathway (e.g. binding reactions) were assumed to be at quasi-equilibrium and thus represented by algebraic equations [210]. As a consequence, we used the IDA solver which is able to handle a system of differential algebraic equations (DAEs). A prerequisite for this solver was that all equations which were solved based on analytical solutions (e.g. CaSS, CaSR and Ca$_i$ and their buffering based on Zeng *et al.* [150]) were reformulated according to the original notation used by Luo and Rudy [149]. Details on this reformulation as well as an explanation of different solver parameters and a first evaluation regarding the solution quality with different termination criteria can be found in [7, 5].

4.4 The Congenital Long-QT Syndrome

As explained in section 2.2.5, the congenital LQT syndrome (LQTS) is one of the most often investigated genetic diseases of the heart. During the course of this thesis, three studies were conducted that investigated several aspects of the disease. The methodology used in these studies will be explained in the following. Each subsection begins with a short description detailing the scope of the investigation.

4.4.1 Multiscale Modeling of Long-QT 2 in the Visible Man Torso: A Feasibility Study

Scope of the Study

The aim of this study was the creation and evaluation of a simplified in-silico model of LQT 2. To this end, a heterogeneous 3D model of the human ventricles was used to simulate action potential propagation and repolarization for LQT 2 and a physiologic control setup. The results of these simulations were used as input for the calculation of the BSPM and ECG. In addition to that, AP morphology traces were extracted from representative transmural locations and the results were compared with experimental [132] and other in-silico studies [133]. The results of this investigation can be found in section 6.4.1.

Methods

In order to simulate the electrophysiological processes in the ventricles and calculate the corresponding ECGs, an isotropic 3D model of the Visible man heart and torso was used in conjunction with the monodomain equation and a model for ventricular electrophysiology [151]. Transmural electrophysiological heterogeneity of I_{Ks} and I_{to} was considered as this has been shown to be important for the realistic simulation of a transmural ECG with a positive T-Wave [301]. Rather than modeling endocardial, M and epicardial cells in discrete, equally thick layers, we chose to calculate the corresponding maximal heterogeneous ion channel conductivities depending on the transmural position using a spline interpolation. The resulting transmural conductivity distribution can be seen in Fig. 4.6. LQT 2 was incorporated into the electrophysiological model by a reduction of the maximum conductance $g_{Kr,max}$ of I_{Kr} [302]. A reduction to 50% $g_{Kr,max}$ was chosen to model a mild version of LQT 2 while a complete block of the current (0% $g_{Kr,max}$) represented a more severe form.

The computational elements which described the ventricular tissue, were coupled with a monodomain approach assuming equal intra- and extracellular conductivities (see section 3.1.3). The corresponding equations were discretized with the finite difference method.

The simulation study was based on the anatomy of the Visible Man dataset, initially provided by the National Library of Medicine, Bethesda, Maryland, USA and post-processed at the Institute of Biomedical Engineering (Karlsruhe Institute of Technology) [303]. The original dataset of the ventricles was interpolated to a cubic voxel size of 0.4 mm (which lead to a model that contained 279 x 254 x 259 voxels). In order to ensure a realistic sequence of ventricular activation, we adapted a Purkinje fiber network that was presented in [185]. In this case, the automatically generated Purkinje fiber endings (see Fig. 4.7A) determined the sites at which we applied endocardial stimulation currents. The sequence of these stim-

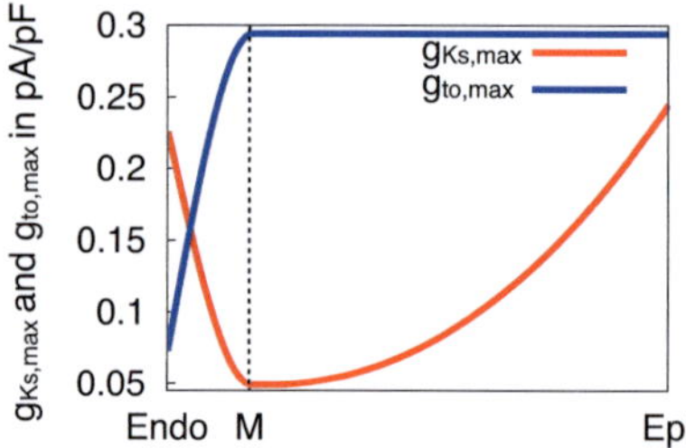

Fig. 4.6. The transmural course of the maximal conductivities $g_{Ks,max}$ and $g_{to,max}$ is shown. In case of $g_{Ks,max}$, the conductivity was lowest in M cells while $g_{to,max}$ was small in endocardial cells.

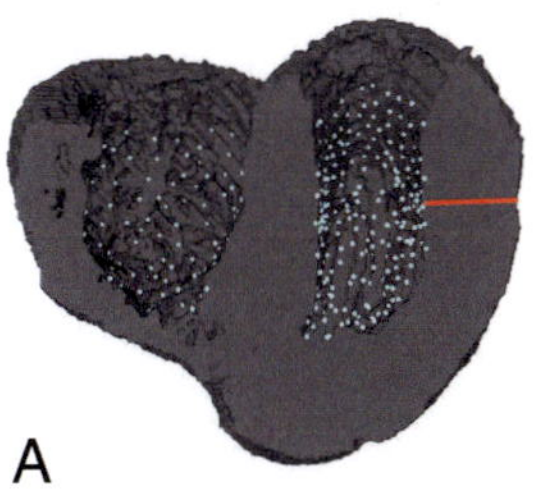

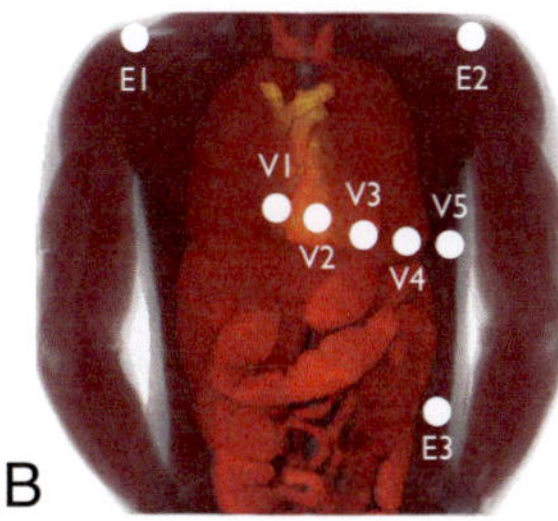

Fig. 4.7. A: Ventricular anatomy of the Visible Man dataset (sliced view). The turquoise dots mark the locations of the Purkinje fiber endings. Stimulus currents were applied at these locations in a special sequence that was designed to mimic the excitation conduction system. Exemplary endocardial, M and epicardial cells were monitored on the red line to analyze the APD changes (while electrotonical coupling was present) that were induced by the LQT 2 syndrome (see section 6.4.1 for details). B: Transparent visualization of the Visible Man torso model with superposed ECG lead positions. The Central Terminal for the Wilson leads was generated using the points E1–E3. Figure adopted from [8].

ulus currents was derived by calculating the distance of each Purkinje fiber ending to a virtual AV node and assuming a constant conduction velocity. Before running the action potential propagation simulations, we pre-calculated the electrophysiological model in an uncoupled environment (25 s with a basic cycle length of 1 s) to tune the gating variables and ionic concentrations.

The torso model which was used during the forward calculations had a relatively coarse grid size of 2 mm (in total it contained 297 x 170 x 260 cubic voxels) to save computation time and memory consumption. It comprised the following list of tissues and organs: lungs, liver, intestine, pancreas, spleen, kidneys, bones, blood, muscle and fat tissue. It is important to consider the respective tissue conductivities for each organ (for details see section 3.2.3.2) during the solution of the forward problem. In this case, tissue conductivities were assigned based on the published values from Gabriel *et al.* [245]. No anisotropic tissue properties were considered in the torso model. Before solving the forward problem, the voxelized torso model was converted into a tetrahedron mesh (for details, see section 3.2.3). Then the simulated transmembrane voltages were interpolated onto the tetrahedron mesh of the ventricles and the body surface potential map was calculated as described in section 3.2.3. The electrode positions that were used for the extraction of the standard Einthoven and Wilson leads can be seen in Fig. 4.7B.

The simulation of action potential propagation and repolarization in the ventricles was conducted on 5 Apple XServe G5 dual 2 GHz processor cluster nodes using the message passing interface (MPI) [304] for process communication. In total, we simulated 450 ms for each of the three electrophysiological setups (physiological case, mild LQT 2: 50% $g_{Kr,max}$, severe LQT 2: 0% $g_{Kr,max}$). Using a 10 μs Euler integration times step, the simulation of each setup took 45 hours. In contrast to that, the solution of the forward problem was far less demanding. The associated equations were solved in 1.5 hours on only one 2 GHz G5 processor (temporal increment: 1 ms).

4.4.2 Suitability of Different Electrophysiological Models to characterize Long-QT 2

Scope of the Study

The aim of this study was to evaluate a set of state of the art electrophysiological models regarding their suitability to model LQT 2. The criteria that were used for the evaluation were on the one hand based on APD changes in a single cell environment and on the other hand on QT duration and T-Wave morphology that were extracted from a computed transmural ECG. The results of this investigation can be found in section 6.4.2.

Methods

Transmural heterogeneity was considered in the electrophysiological models by varying the maximum channel conductivity of I_{Ks} ($g_{Ks,max} = g_{Ks,Epi}$) and I_{to} ($g_{to,max} = g_{to,Epi}$). In this case, epicardial conductivities were adopted from the original publications [151, 152, 305, 306] while the endocardial and M cell conductivities were adapted as follows:

- Concerning I_{Ks}, we set M cell conductivity to $g_{Ks,M} = 0.2 \cdot g_{Ks,Epi}$ and endocardial conductivity to $g_{Ks,Endo} = 0.92 \cdot g_{Ks,Epi}$.

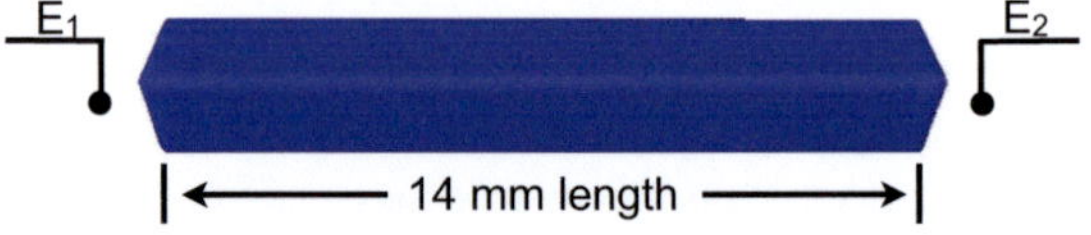

Fig. 4.8. The in-silico model of the parallelepiped (blue color) was 14 mm long and placed in a bath-medium with electrodes at either side to extract the extracellular potentials. The potential difference between the electrodes determined the course of the transmural ECG.

- Regarding I_{to}, we set M cell conductivity to $g_{to,M} = g_{to,Epi}$ and endocardial conductivity to $g_{to,Endo} = 0.25 \cdot g_{to,Epi}$.

Conductivity values in between these anchor points were interpolated using a cubic spline. This resulted in a similar transmural conductivity distribution as shown in Fig. 4.6.

LQT 2 was modeled by reducing the maximum conductivity $g_{Kr,max}$ by 50% to account for the LQT 2 induced loss of function of I_{Kr}. All electrophysiological models were pre-calculated (25 s of precalculation with a basic cycle length of 1 s) to tune gating variables and ionic concentrations before the APD was extracted for subsequent evaluation.

In order to investigate the impact of LQT 2 on the transmural ECG, we simulated excitation spread and repolarization in a synthetically generated isotropic parallelepipedal structure consisting out of 70 cubic voxels with a side length of 0.2 mm. The coupling of the ventricular tissue was described based on the bidomain equations (see section 3.1.3) which directly allowed the extraction of the extracellular potentials Φ_e (see Fig. 4.8) without the need for an additional forward calculation procedure.

A number of characteristic signal features was extracted from the transmural ECG in an effort to characterize the morphological changes that were introduced by LQT 2. Among them was the QT interval which was defined here as the time between QRS onset and the point at which the tangent drawn at the maximal downslope of the T-Wave intersected with the isoelectric line (see Fig. 4.9). In addition to that, the full width half maximum of the T-Wave (T_{FWHM}) was used as a measure for the T-Wave shape. Finally, the ratio T_{max}/QRS_{mean} was an indicator for the height of the T-Wave relative to the QRS complex. Both T_{FWHM} and T_{max}/QRS_{mean} were modulated by the transmural dispersion of repolarization.

4.4.3 In-silico Evaluation of Beta-Adrenergic Effects on the Long-QT Syndrome

Scope of the Study

As explained in section 3.1.6, several in-silico studies tried to reproduce the wedge experiments from Antzelevitch's group in which LQT was induced pharmacologically and the T-Wave changes were monitored in the transmural ECG. However, these previous in-silico studies were not able to consider the effects of beta-

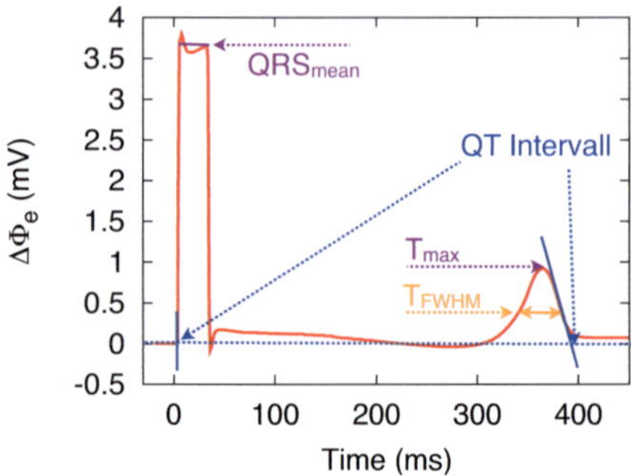

Fig. 4.9. Exemplary transmural ECG together with the parameters that were extracted in an effort to characterize the morphological changes introduced by LQT 2. The QT interval was determined based on the tangent method. T_{FWHM} and T_{max}/QRS_{mean} were used to quantify changes of the T-Wave concerning amplitude and signal morphology.

adrenergic regulation as the underlying effects were not known in detail and thus a corresponding model was not available at that time.

Recently, such a model was presented by Saucerman et al. [100] and we subsequently integrated it into the model of human ventricular electrophysiology from ten Tusscher et al. (see section 4.3). In this study, we used this expanded model to mimic the wedge experiment from Antzelevitch *et al.* (see section 3.1.6). A special focus was put on changes in TDR in each of the three different LQT subtypes (LQT 1-3) both in the presence and absence of beta-adrenergic stimulation. The results of this study can be found in section 6.4.3.

Methods

The computational model of the ventricular wedge preparation contained 120 x 20 x 20 cubic voxels with a side length of 0.1 mm. Thus the resulting transmural extent of the wedge was comparable to experimental preparations [221, 217]. The anisotropic properties of the ventricular tissue were considered by creating a fiber orientation setup for the in-silico wedge model as described in section 3.2.2.

Ventricular electrophysiology was described based on the revised version of the ten Tusscher model [152] in which we integrated the beta-adrenergic signaling cascade from Saucerman *et al.* [7] (for details please refer to section 4.3). The model was recently expanded to consider adrenergic effects on the Na/K-ATPase (I_{NaK}). These effects help to remove excessive Ca^{2+} from the cytosol which could otherwise trigger EADs by inducing spontaneous Ca^{2+} sparks from the sarcoplas-

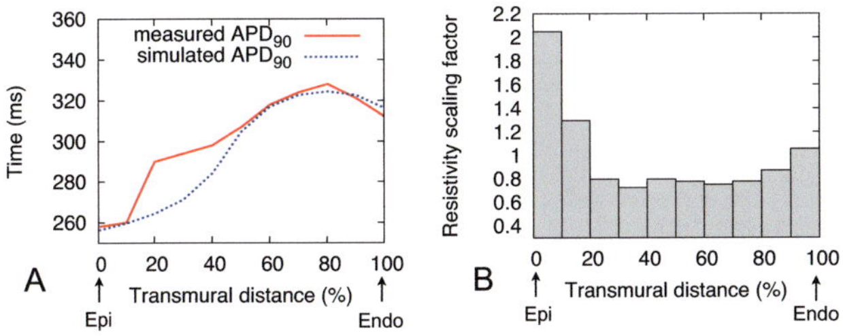

Fig. 4.10. A: Measured [119] and simulated transmural APD_{90} distribution for the wild-type setup. A constant offset of 35 ms was added to the measured data to compensate for the shorter APD_{90} of the canine compared to human myocytes. Thus a better comparability between the measured and simulated APD_{90} trace was ensured. B: Resistivity scaling factors were extracted from [119] and used to adapt the transmural conductivities in our model.

mic reticulum [299] (for details please refer to section 6.3.1). Furthermore, we adapted the distribution of the transmural electrophysiological heterogeneity and the maximum conductivity $g_{Ks,max}$ in each of the three transmural layers such, that we received a similar ratio of $APD_{90,Epi}/APD_{90,M}/APD_{90,Endo}$ as in the wedge experiments [119]. As no information was provided on the transmural distribution and width of the endocardial, M and epicardial layer we assumed the following distribution: 20% endocardium, 30% M and 50% epicardium. Likewise, we adapted $g_{Ks,max}$ by setting $g_{Ks,Endo} = 0.45$ ns/pF, $g_{Ks,M} = 0.08$ ns/pF, $g_{Ks,Epi} = 0.75$ ns/pF. Finally a non-uniform distribution of tissue resistivity was reported throughout the wall of the wedge preparation [119]. To incorporate this data into our model, we extracted resistivity scaling factors from the measurements and used them to adapt our conductivities in 10 transmural layers accordingly. The resulting transmural APD_{90}-distribution and tissue resistivities between epicardium and endocardium can be seen in Fig. 4.10.

The wedge was electrically activated by applying intracellular stimulus currents at its endocardial front surface (see Fig. 4.11). The bidomain model was used to describe action potential propagation within the wedge (see section 3.1.3). In order to calculate a transmural pseudo ECG (tECG), we submerged the wedge in blood ($\sigma_{blood} = 0.7$ S/m) and placed a ground electrode near the endocardial surface. The extracellular potentials were then extracted in 1 mm distance from the epicardial surface and saved as tECG.

In order to model the effects of LQT 1 and LQT 2, we used a simple model which was based on a reduction (50%) of the maximum conductivity $g_{Ks,max}$ and

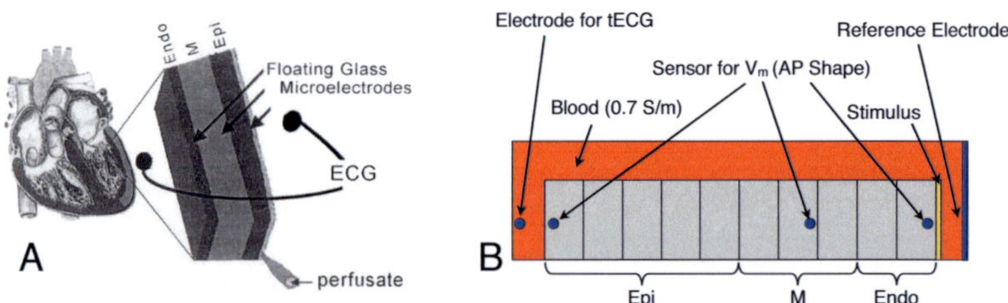

Fig. 4.11. A: Wedge preparation according to [262]. B: In-silico model of the wedge experiment from Yan *et al.* [262]. The wedge was submerged in blood and sensors were used to extract the LQT-induced changes in AP morphology and the transmural pseudo ECG (tECG).

$g_{Kr,max}$, respectively. In contrast to that, LQT 3 was characterized by a more complex Markov model of I_{Na} which described a mutation in the C terminus of the channel (1795insD) according to [307]. The beta-adrenergic signaling cascade was activated by applying $1\,\mu M$ of the adrenergic agonist isoproterenol (ISO). Before action potential propagation was simulated in the wedge, all parameter configurations of the electrophysiological model were pre-calculated in an uncoupled environment for 120 s with a basic cycle length of 2000 ms (same basic cycle length as in the experiments [221]) to adjust gating variables and ionic concentrations and to ensure that the adrenergic effects are fully present.

5

Anatomical Modeling and the Forward Problem of Electrocardiography

This chapter introduces the methods used for the various studies in the realm of *anatomical modeling and the solution of the forward problem* that were conducted during the course of this thesis (i.e. the creation of anatomical models and effects of model resolution on the electrophysiological simulations, investigations on the role of skeletal muscle fiber orientation in the context of the forward problem, the modeling of ventricular contraction and its effects on the ECG, the ranking of tissue conductivities for a realistic simulation of ECG signals and a novel PCA based BSPM prediction technique). The corresponding results can be found in chapter 7.

5.1 Creating Anatomical Models based on MRI data

For the realistic simulation of cardiac electrophysiology and the calculation of the corresponding body surface ECGs it is important to have accurate anatomical models that deliver geometrical boundary conditions. In the past, most studies were based on standard datasets like the Utah torso model [243], NORMAN [308] or the Visible Man [309, 303] and Visible Female [310, 311] dataset. However, over the last years, patient-specific anatomical models are becoming more and more popular. This has two main reasons:

- The anatomy influences both the action potential propagation (e.g. due to differences in size of the ventricles, thickness of walls, possible obstacles) as well as the solution of the forward problem (e.g. due to differences in shape, size and position of the organs, torso contour). As cardiac modeling should be transferred to the clinical practice, future simulations will have to be performed on patient-specific models and the sensitivity of these simulations with respect to certain anatomical features will have to be investigated.

- It is difficult to validate the simulations based on a standard model as the necessary data (e.g. intracardiac measurements, multi-channel ECG recordings) is usually not available. This is different for patient-specific datasets where electrophysiological mapping data or ECGs can be acquired alongside the anatomical imaging.

Both, CT and MRI deliver image data that allow to create anatomical models of sufficient detail and resolution for most applications of cardiac modeling. However, as this study was based on data from volunteers (probands/patiens) only MR images were available due to ethical reasons.

After the image acquisition was completed, the data had to be segmented and classified in order to be able to assign the corresponding electrical properties to the different tissues. Segmentation can be done manually or (semi-)automatically. Although both approaches have their advantages and disadvantages, automatic methods will dominate in the clinical practice (in the future) as manual segmentation is very time-consuming and requires expert knowledge. Automatic segmentation procedures are often based on pre-defined mean models that are adapted during the segmentation procedure [312, 313]. Such techniques have also been used at the Institute of Biomedical Engineering (Karlsruhe Institute of Technology) by Krueger *et al.* [314, 315, 316]. However, they were not yet available during the course of this work. Here, segmentation techniques like region growing or interactively deformable triangle meshes were used. Both segmentation techniques were previously implemented at the Institute of Biomedical Engineering [317, 318, 319, 320, 321] and were merely used in this work for the creation of patient-specific models.

5.1.1 Region Growing

The region growing algorithm is usually initialized using manually chosen seed points. Starting from these seed points, the algorithm compares the gray values of neighboring voxels with a pre-defined range of accepted values. If the gray value is found within this range, the voxel will be marked and subsequently included in the segmentation results. This newly added voxel acts as additional seed point from which further comparisons will be conducted. In theory, the algorithm should stop at organ borders when no new neighbors with a matching set of features can be found. An advantage of the method is, that the resulting regions are always homogeneous (e.g. compared to thresholding). However, on the downside, the segmented contours can be rough and chiseled which is problematic if a tetrahedron mesh should be created based on the segmentation results. In general, the results of the region growing algorithm depend on the chosen seed points. However, recently an extension has been proposed that overcomes this limitation (although this extension was not used in the presented work) [322]. In this work, region growing was mainly used for the segmentation of the intestine or blood pools. Fig. 5.1 shows a schematic representation of the region growing algorithm.

5.1.2 Interactively Deformable Triangle Meshes

Interactively deformable triangle meshes are a manual version of the so-called active contours [323]. The segmentation procedure is initialized by superposing a spherical triangle mesh over the MR image data. The user is now able to manipulate the contour of this mesh in an effort to approximate the borders of the organ that should be segmented. To accelerate the segmentation, it is usually possible to adjust the area of effect of the mesh manipulation force. A large area of effect will lead to more global changes (nodes in a relatively large neighborhood of the chosen node will be moved) whereas a small area of effect limits the deformation to the direct vicinity of the chosen node. If the area of effect can be chosen, it is advisable to start the segmentation by roughly deforming the mesh to the target organ (large area of effect) before the details are traced with a higher precision (small area of effect). The use of interactively deformable meshes results in homogeneous regions that normally have smooth surfaces. Furthermore triangle meshes are very

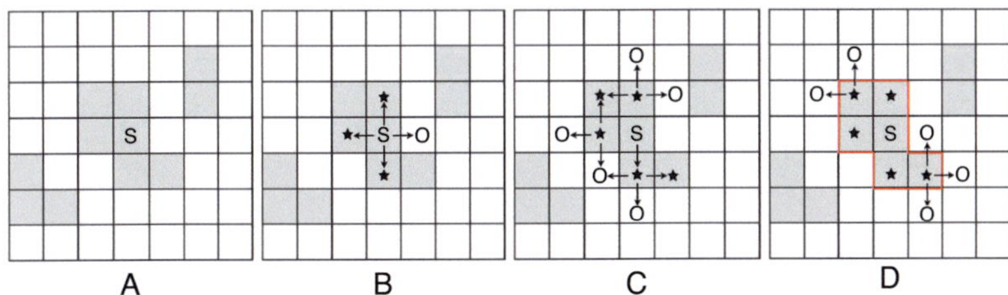

Fig. 5.1. Schematic visualization of the region growing technique. A: A seed point is manually placed in the structure that should be segmented. B: Starting from this seed point, the gray value of the neighboring voxels is compared with a pre-defined range of accepted values. Voxels that have gray values within this range are included in the region and marked with a star, whereas all other voxels are marked with an "O" (outside). C: The newly added voxels act as seed points and the algorithm iterates. D: The algorithm stops as soon as there are no new neighbors in the target gray value range. The red contour marks the segmented region.

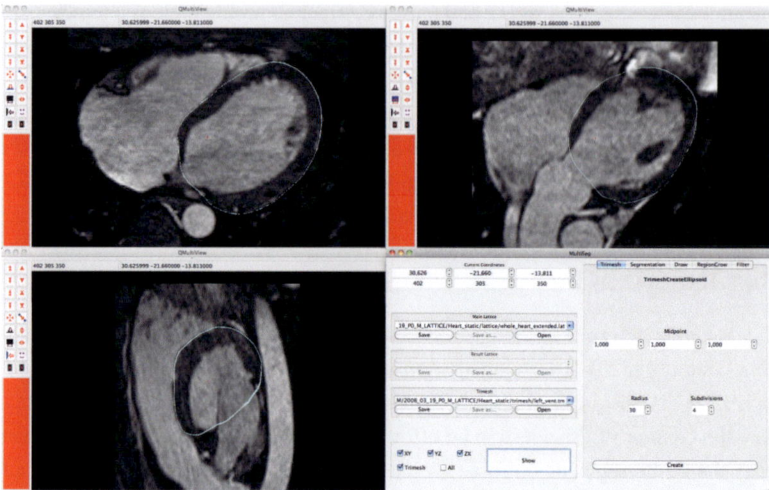

Fig. 5.2. Graphical user interface that allows the manipulation of the interactively deformable triangle meshes. In this case, the triangle mesh is displayed in turquoise on top of the MRI data of the heart. The MRI data is visualized from three different perspectives (XY, XZ, YZ plane) to facilitate the navigation and the segmentation of the data. Initial center and radius of the spherical triangle mesh as well as the area of effect of the manipulation force can be set in the window on the lower right side.

flexible and allow the segmentation of low quality image data where automatic methods would be unsuitable (in this case, the experience of the user can compensate for poor image quality). However, their use requires expert knowledge, results are user dependent and the segmentation procedure is very time consuming. In this work, all structures in the ventricles and torso (exceptions: see 5.1.1) were segmented using this technique. A visualization of the graphical user interface that allowed the manipulation of the interactively deformable triangle meshes can be seen in Fig. 5.2.

5.2 Resolution Effects of Anatomical Models

As explained in section 3.2.1 the spatial resolution of the anatomical model can influence the conduction velocity and wavefront shape during simulations of action potential propagation in tissue. In order to quantify these effects we created virtual tissue patches with differing resolutions and compared the associated isochrone distributions. The results of this in-silico experiment were then used to choose an adequate resolution for the simulation of a realistic excitation sequence in an anisotropic and electrophysiologically heterogeneous biventricular model. The methods of this investigation will be presented below while the results can be found in section 7.2.

5.2.1 Investigations on Wavefront Shape

In order to investigate the effects of spatial resolution on the wavefront shape, we created three virtual tissue patches with the dimensions 6 cm x 6 cm (see Fig. 5.3). Each patch consisted of cubic voxels with different resolutions. For voxel side lengths of 0.4 mm, 0.2 mm and 0.1 mm, the patch dimensions translated into datasets containing 150 x 150, 300 x 300 and 600 x 600 elements, respectively. The homogeneous electrophysiological properties of the tissue patches were described using the epicardial model for human ventricular myocytes from ten Tusscher *et al.* [151]. As indicated by the black arrows in Fig. 5.3, the fiber orientation was set to $0°$. Activation was initiated by applying a stimulus current to the center of each tissue patch. The conduction velocity in both transverse and longitudinal fiber direction was identical in all three patches (transverse velocity: 50 cm/s; longitudinal velocity 65 cm/s according to [324]). To enable this identical conduction velocity, we had to adapt the intracellular transverse and longitudinal conductivities as the conduction velocity is known to be reduced at lower spatial resolutions [166]. Adapted intracellular conductivities were set to the following values:

- 0.4 mm: $\sigma_{i,T} = 0.095$ S/m and $\sigma_{i,L} = 0.141$ S/m
- 0.2 mm: $\sigma_{i,T} = 0.072$ S/m and $\sigma_{i,L} = 0.114$ S/m
- 0.1 mm: $\sigma_{i,T} = 0.063$ S/m and $\sigma_{i,L} = 0.102$ S/m

5.2.2 Biventricular Model

The anatomical information for the construction of the biventricular model was derived from the MRI scans of a healthy volunteer. The segmented voxelized dataset of the ventricles consisted of 704 x 492 x 550 ($\approx 191 \cdot 10^6$) elements with a side length of 0.2 mm. Approximately 15.5% of the dataset was excitable ventricular tissue ($\approx 30 \cdot 10^6$ voxels). Fiber orientation was incorporated into the model as described in section 3.2.2 using a rule-based approach. Similar as in the tissue patch experiments described above, we used the ten Tusscher model for human ventricular myocytes [151] to describe the electrophysiological properties of the tissue. In

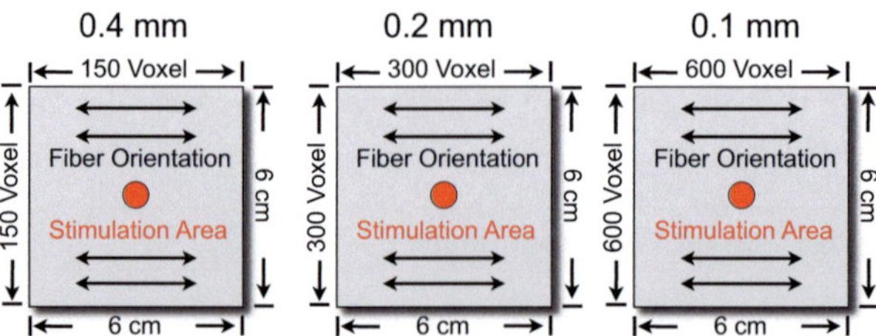

Fig. 5.3. Schematic representation of the three virtual patches that were used to investigate the effects of spatial resolution on the wavefront shape. Lowest resolution was 0.4 mm. In this case, the tissue patch consisted out of 150 x 150 cubic voxel elements. In contrast to that, the tissue patch with the highest resolution contained 600 x 600 voxels (resolution: 0.1 mm).

this case, however, electrophysiological heterogeneities were considered to ensure a realistic repolarization sequence. To include transmural heterogeneities, the wall was divided in three layers of different thicknesses: 40% endocardium, 40% M cells and 20% epicardium. In each of these layers, the parameters g_{Ks} and g_{to} were modified according to the suggestions from ten Tusscher et al. [151]. In addition to these transmural electrophysiological heterogeneities, we adapted the tissue conductivities in ten distinct transmural zones. Conductivity scaling factors for each of these zones were derived from the wedge measurements from Yan *et al.* [262]. The setup of electrophysiological heterogeneities was completed by the inclusion of an apico-basal gradient of g_{Ks}. Based on Szentadrassy *et al.*, apical g_{Ks} was doubled compared to the base [263]. All values in between apex and base were linearly interpolated. The excitation conduction system and in particular, the Purkinje fiber network was modeled by a special sequence of endocardially applied stimulations (see [2] and section 4.1).

5.2.3 Numerical Methods

The mathematical equations of the ten Tusscher model were solved with a combination of the Euler and Rush-Larsen [267] method while the action potential propagation in the tissue was described with the monodomain reaction diffusion model (see section 3.1.3). Temporal increments were set to 10 μs for the electrophysiological model as well as for the monodomain equation in case of the three tissue patches and to 20 μs in case of the biventricular simulations. All simulations were performed in a C++ simulation framework [325].

5.3 Modeling Skeletal Muscle Fiber Orientation and the Effects on the Body Surface ECG

As explained in section 3.2.3.3 skeletal muscle fiber orientation has a strong effect on forward calculated ECGs. Yet it is currently not feasible to image the fiber arrangement non-invasively, which is a prerequisite to be able to include it in patient-specific torso models. Therefore only simplified rule-based approaches can be used to consider these anisotropic properties during the forward calculation.

In this study, we evaluate the realism of several rule-based methods. Among them is the approach from Klepfer *et al.* and two newly developed methods which are based on Klepfer's assumptions of tangential orientation of the fibers with respect to the torso's surface. The quality of the different fiber orientation approximations was assessed by comparing the associated forward calculated BSPMs with the BSPM resulting from the reference skeletal muscle fiber distribution developed by Sachse *et al.* for the Visible Man dataset [253].

The methods of this investigation will be presented below while the results can be found in section 7.3.

5.3.1 Data Source: Anatomical and Electrophysiological Model

As the reference fiber orientation was derived from the Visible Man dataset, it had to provide the anatomical basis for the simulation of cardiac activation and repolarization as well as for the calculation of the body surface ECGs. Action potential propagation was calculated in an anisotropic and electrophysiologically heterogeneous model of the ventricles. Details on the considered cardiac anisotropy information and on the distribution of transmural and apico-basal heterogeneities can be found in [12, 2].

In order to solve the forward problem, the simulated transmembrane voltages were interpolated onto an unstructured tetrahedron mesh of the ventricles and the BSPMs were calculated as described in [15]. Fig. 5.4A shows a visualization of the voxelized torso model that was used as basis for the construction of the unstructured tetrahedron mesh. It contained 18 different tissues and fluids among which were: blood, lungs, liver, intestine, pancreas, spleen, kidneys, muscle (skeletal and heart), bones, cartilage and fat tissue. Anisotropy of electrical conductivity was considered in case of the ventricles and skeletal muscle. Both anisotropy ratios (along:across) were set to 3:1. The tissue conductivities used for the forward calculations were based on the measurements from Gabriel *et al.* at 10 Hz [245].

Electrode locations which were used to extract the body surface potentials are visualized in Fig. 5.4B.

5.3.2 Creation of Different Datasets Containing Skeletal Muscle Fiber Orientation

The basic assumption of all rule-based approaches was that the skeletal muscle fibers are aligned parallel to the torso surface. Furthermore, all existing rule-based methods neglect the muscle fiber component that is oriented from head to feet (longitudinal orientation). In this evaluation, all skeletal muscle fiber information was integrated into the skeletal muscle layer of the Visible Man dataset that is displayed in Fig. 5.5A. We now list the fiber orientation datasets that were considered in this evaluation:

- Gold: This dataset was used as reference (gold standard). It was created based on the thin-section photos of the Visible Man dataset [253]. Automatic methods such as texture analysis and a 3D Sobel filter were used to derive an initial orientation which was then manually revised by human experts.
- Klepfer: In this case, the fiber orientation was assumed to be perpendicular to the bisector of the twelve segments into which the thorax was subdivided [243]. A visualization of this method can be found in Fig. 5.5B.
- Gradient: Here, a 3D Sobel filter was used on the torso geometry filled with an increasing gray value from inside to outside (see Fig. 5.5C). The result was a

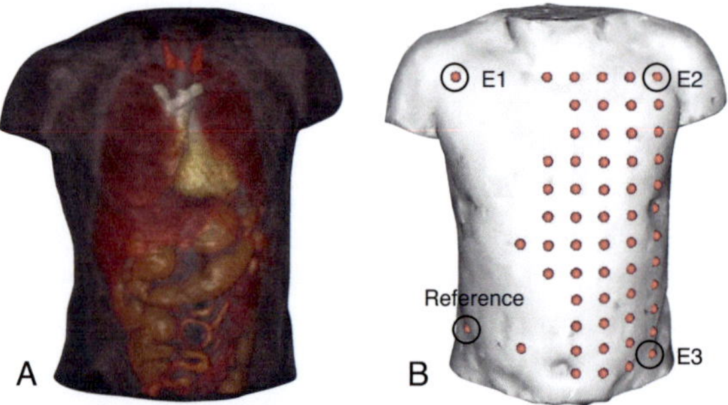

Fig. 5.4. A: Voxelized model of the Visible Man torso. In total it contained 18 different tissues and fluids. The most important were blood, lungs, liver, intestine, pancreas, spleen, kidneys, muscle (skeletal and heart), bones, cartilage and fat tissue. B: Unstructured tetrahedron model of the Visible Man torso which was created based on the voxelized model shown in (A) and used for the forward calculations. The electrode positions mark the locations were the body surface potentials were extracted (7 electrodes were located on the back). Einthoven I: E2-E1, Einthoven II: E3-E1.

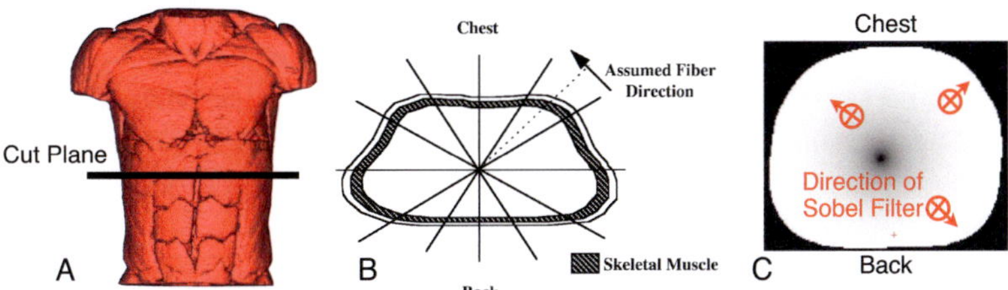

Fig. 5.5. A: Visualization of the skeletal muscle layer that was part of the Visible Man model. All rule-based fiber orientation information was integrated into this layer. The cut plane indicates the viewpoint in (B) and (C). B: Rule-based approach according to Klepfer *et al.* [243] (Figure directly adopted from [243]). The thorax was subdivided into twelve segments. The fiber orientation was assumed to be perpendicular to the bisectors of each segment. C: The torso was filled with an increasing gray value from inside to outside. A 3D Sobel filter used on this dataset created a vector field in which each vector was normal to the thoracic surface at its respective location. The fiber orientation was now assumed to be perpendicular to the planes formed by the normal vectors and the longitudinal torso orientation.

vector field in which each vector was normal to the thoracic surface at its respective location. The fiber orientation was now assumed to be perpendicular to the planes formed by the normal vectors and the longitudinal torso orientation.

- Gold+No-Z: In this case, we used the gold standard fiber orientation. However, the longitudinal component of the fiber vectors was erased so that all fibers were horizontally aligned.

- Gradient+Back: This setup was equal to the Gradient dataset. The only exception was the incorporation of the back muscles. They are known to be longitudinally oriented and were integrated accordingly.

- Only-Heart: No skeletal muscle anisotropy was considered (only ventricular fiber orientation data was taken into account).

5.4 Modeling Ventricular Deformation and the Effects on the Body Surface ECG

In section 3.2.3.4 we explained that the forward problem is usually solved using a static heart and torso model that neglects the movement of the cardiac sources during ventricular contraction and relaxation.

In this study, we evaluate the effects of this movement on the BSPM by creating an anisotropic and electrophysiologically detailed *dynamic* model of the ventricles based on cinematographic (Cine) and tagged MRI data. As the ventricular contraction and relaxation occurs during the ST segment and T-Wave of the ECG, we hypothesized that the magnitude of T-Wave changes is likely to depend on the relation between mechanical deformation and electrical repolarization. In other words, the synchrony between maximal systolic contraction and onset of ventricular repolarization will probably determine the extent of the BSPM changes.

We tested this hypothesis by creating three different electrophysiological setups, each with a unique QT time:

- The deformation-induced BSPM changes for a healthy volunteer were assessed by using a setup with physiological QT time
- Effects for an abnormally short QT time (fast repolarization) were investigated based on a model of the short QT syndrome (SQT) (mechanical contraction and relaxation was identical as in the physiological setup)
- Effects for an abnormally prolonged QT interval (slow repolarization) were evaluated using a model of the long QT syndrome (LQT) (mechanical contraction and relaxation was identical as in the physiological setup)

Both, SQT and LQT are generally assumed to be "primary electrical diseases" with no influence on mechanical function [137]. They should therefore be suitable to probe to which extent the T-Wave changes depend the synchrony/asynchrony of mechanical relaxation and electrical repolarization.

5.4.1 Rationale of the Construction of the Dynamic Model

Fig. 5.6 outlines the main steps that led to the construction of the dynamic forward calculation model. Details on the different steps will be provided in the subsequent sections.

At first, time dependent anatomical models of the ventricles were constructed based on the 4D Cine MRI scans. These models captured the mechanical states of the ventricles during contraction and relaxation. They were also used as references for the elastic registration procedures during which the diastolic dataset was mapped onto the various contracted states. The resulting displacement information was used to move the nodes of the unstructured tetrahedron mesh (used for the forward calculation) to their corresponding positions in the dynamic model.

Then the electrical activation and repolarization was calculated on the static diastolic model and the electrical sources were moved to their positions in the dynamic model prior to the forward calculation. The limitations that were associated with this hybrid approach will be discussed in section 7.4.4.

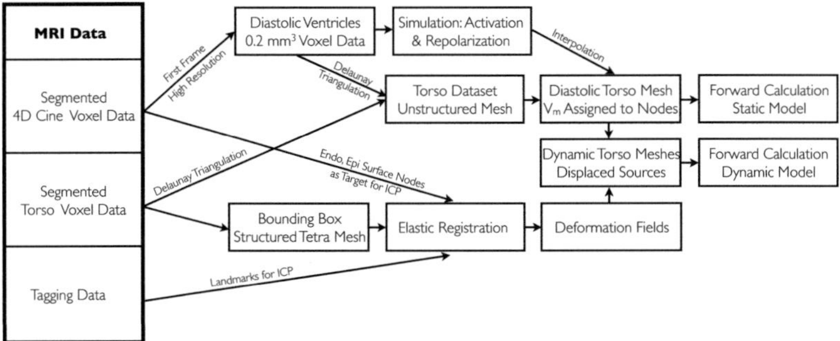

Fig. 5.6. The flowchart provides an overview over the main steps that led to the construction of the dynamic model. The first frame (diastolic state) of the 4D Cine data was used to simulate the depolarization and repolarization of the ventricles. The resulting transmembrane voltage distributions were assigned to the nodes of an unstructured tetrahedron mesh, generated from the torso voxel data and the high-resolution diastolic heart model. This static mesh was used for the forward calculation in case of the static model. In order to conduct dynamic forward calculations, the diastolic model of the ventricles was registered to the anatomical models in deformed states that were created based on the remaining Cine data. The result of each registration was a deformation field which was used to move the nodes and corresponding transmembrane voltages to their position in the dynamic model.

5.4.2 Anatomical Modeling

5.4.2.1 MR Imaging protocol

A healthy, 27-year-old volunteer delivered the MR images that formed the anatomical basis of this study. MR image acquisition was performed on a clinical 1.5T scanner (Magnetom Avanto, Siemens Medical Systems, Erlangen, Germany). Thorax imaging was based on a T1 weighted 3D gradient echo sequence (VIBE - volumetric interpolated breath-hold sequence) in an expiratory breath-hold (parameters were: coronal orientation, voxel size 1 x 1 x 2 mm^3, TR/TE: 3.2/1.1 ms, FA: 8°) [326].

In contrast to that, three different datasets of the cardiac cavities were acquired:

- 22 short-axis slices containing Cine MRI data which was preprocessed and segmented in different contraction states in order to be used as references for the registration procedures.

- A high resolution diastolic dataset which was used to validate the diastolic segmentation based on the lower-resolution Cine data.

- 3 slices containing tagging data which was used in one case during the registration to incorporate information concerning the regional heterogeneity of deformation.

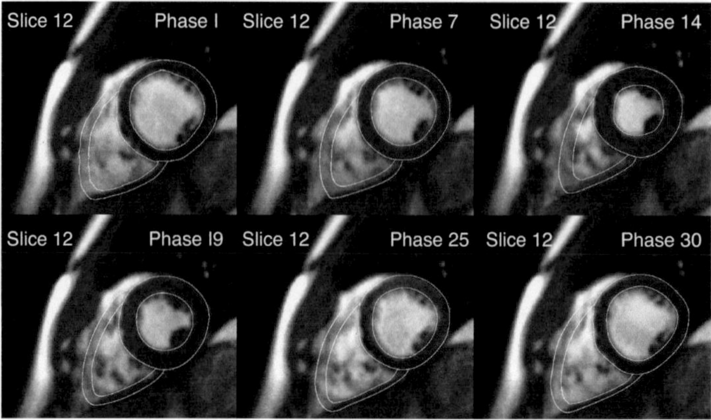

Fig. 5.7. This figure shows one of the 22 short axis slices that contained the Cine data. Each Cine slice imaged the ventricular contraction and relaxation in 50 phases (RR-Interval: 1060 ms, temporal resolution: 21.6 ms). Shown are selected phases of slice 12. Phase I shows the diastolic state, whereas phase 14 coincides with the maximal contraction during systole. Only the first 30 phases were considered during the construction of the dynamic model.

In case of the Cine data, the left and right ventricle were imaged from base to apex in short axis view with a retrospective ECG gated 2D steady-state free-precession sequence (SSFP, trueFISP) Cine (TE/TR: 1.19/35.62 ms, TR-real (echo-spacing): 2.7 ms, : FA 80°, slice thickness: 4 mm, in-plane resolution: 1.9 x 1.9 mm^2). The temporal resolution was 21.6 ms (RR-Interval: 1060 ms, 50 phases). An example of a short axis Cine dataset can be seen in Fig. 5.7. Shown are selected phases from slice 12.

The high resolution diastolic dataset (static) was imaged with a standard 3D ECG and respiratory gated SSFP sequence ("whole heart approach") with a spatial resolution of 1 x 1 x 1 mm^3 [327]. The T1/T2 contrast was chosen such that a good differentiation between the blood filled cardiac chambers and the myocardium could be obtained.

The tagging sequence was based on a segmented FLASH (Fast Low Angle SHot) 2D sequence with prospective ECG gating. Using this technique, a 8 mm grid was "tagged" onto the myocardium (other sequence parameters were: TR/TE: 40.85/3.93 ms; no iPAT, flip angle 14°, spatial resolution 1.3 x 1.3 x 6 mm^3). Overall 22 phases were acquired.

5.4.2.2 Segmentation of the Torso and Cardiac Cavities

In a preprocessing step, all MRI image data were trilinearly interpolated to ensure an adequate isotropic resolution for the segmentation procedure. Then the 22 short axis slices that contained the Cine images were concatenated such, that we received a complete 3D dataset of both left and right ventricle for each of the 50 phases. An example of this fusion of image data is displayed in Fig. 5.8. In this case, the different short axis images are combined to create a 3D dataset of phase 1 (diastolic state).

The aim of this project was the evaluation of the impact of ventricular contraction on the T-Wave not only for a physiological relation between electrical repolarization and mechanical relaxation but also for two pathologies that alter the onset of electrical repolarization. Thus, it had to be ensured that the dynamic model was able to describe ventricular contraction until electrical repolarization was completed in all three electrophysiological configurations (physiological QT time, SQT and LQT). In this case, the LQT setup had the longest QT time ($QT_c >$ 450 ms) thereby determining the minimal temporal extent of the dynamic model. Due to the temporal spacing of 21.6 ms between the different phases in the Cine images (see 5.4.2.1) it was decided to segment the first 30 out of 50 phases. These first 30 phases allowed the description of the ventricular contraction and relaxation up to 626 ms after the R peak which was sufficient, even for severe cases of LQT. The segmentation of the ventricles in differently contracted states and the thorax was done manually, relying on interactively deformable 3D contours (for more information on this technique, see 5.1.2). In case of the segmentation of the ventricles, an important feature of these contours was that they allowed incremental changes. This means that the different phases were not independently segmented by starting each segmentation procedure from scratch. We rather segmented consecutive phases by adapting the contours of the preceding phase (e.g. phase 2 was segmented by adapting the contours of phase 1). This guaranteed consistent results and reduced the overall segmentation errors. On the downside, the reuse of the 3D contours could potentially lead to jagged contour surfaces. This effect was compensated by spatially smoothing the contours with an HC Laplacian filter. In contrast to a conventional Laplacian filter, the HC version was able to prevent shrinkage effects during the smoothing [328]. In addition to that, all contours except for the diastolic, systolic and last contour were averaged over time. Temporal averaging was done based on a moving average window that contained 3 contours (phase $t - 1$, phase t, phase $t + 1$). The resulting spatio-temporally averaged contours were again projected onto the MR images and checked for consistency.

Fig. 5.9 shows the segmented diastolic (transparent gray) and systolic (red) dataset from two different perspectives. Due to the incremental segmentation procedure, the fidelity of the diastolic segmentation was especially important as it served as starting basis for the segmentation of the 29 consecutive phases of deformation. Thus it was validated by comparing it to the independently segmented dataset that was derived from the high-resolution static MRI scan (SSFP, "whole heart approach"; see 5.4.2.1).

The segmented biventricular models in different contraction states served as reference for the elastic 3D registration procedure that is described in the section 5.4.3. All electrophysiological simulations (depolarization and repolarization) were conducted in a voxelized model of the diastolic ventricles containing 691 x 489 x 536 cubic elements with a side length of 0.2 mm. Fiber orientation in the ventricles was modeled as described in section 3.2.2. Its inclusion allowed to consider the anisotropic conduction properties of the myocardium. The anisotropy ratio was set to 2.4 and the conductivities were chosen such that we obtained a realistic transverse conduction velocity of 50 cm/s (longitudinal conduction velocity was 82 cm/s) according to [324] and realistic transmural conduction times similar to [262].

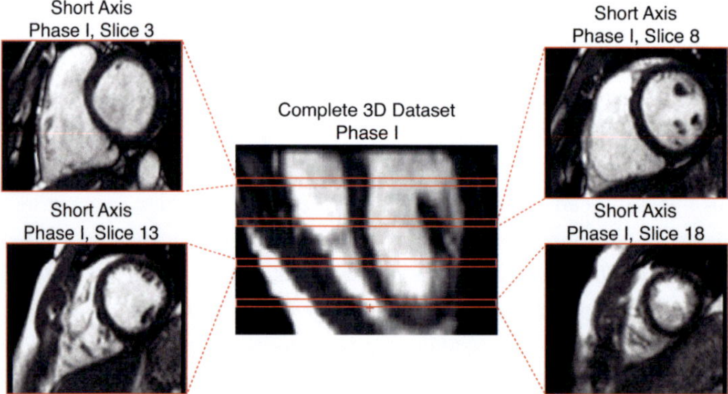

Fig. 5.8. In order to construct a 3D model that represents the deformation state of the ventricles in each phase, the short axis slices were concatenated. Potential misalignment between the short axis slices due to proband movement or heart rate variability that affected ECG gating was corrected manually. The segmentation results of these 3D datasets were subsequently used as input for the registration procedure.

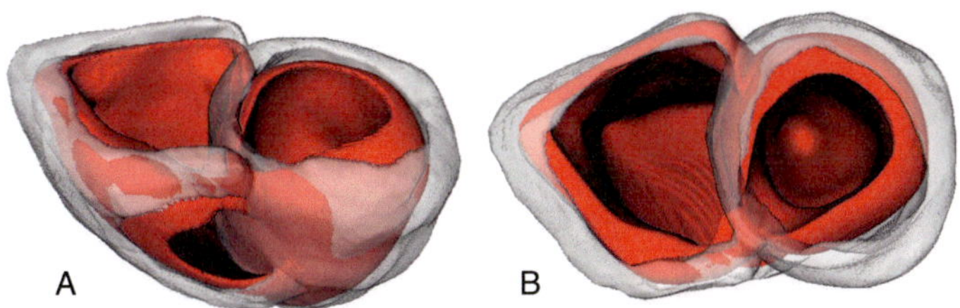

Fig. 5.9. Segmentation results of the diastolic (transparent) and systolic (red) state from side (A) and top (B) view.

Fig. 5.10 shows a voxelized representation of the torso (437 x 500 x 232 elements with a voxel size of 1 mm^3). The final model contained all organs that are relevant for the forward calculation [15] and additional structures like the intestine, spleen and kidneys. The complete list of organs which had distinct conductivities in the model comprised: the lungs, liver, spleen, kidneys, small intestine, colon as well as skeletal muscle, anisotropic heart muscle, blood and fat tissue. The anisotropic tissue properties of the heart muscle were considered during the forward calculation (anisotropy ratio of electrical conductivities was 3:1). Conductivity values were assigned based on the measurements from Gabriel *et al.* at 10 Hz [245].
The forward calculations were conducted on an unstructured tetrahedron mesh that contained a higher node density in the heart than elsewhere. Surface and organ nodes of this tetrahedron mesh were created based on the voxelized torso model shown in Fig. 5.10 while all cardiac nodes were inserted based on the high resolution diastolic dataset. The final tetrahedron mesh that was used for the forward calculations contained 317,000 nodes and 2,040,000 tetrahedrons. For validation purposes, experiments were conducted with meshes that had up to 1,680,000 nodes and 10,400,000 tetrahedrons. However, the results were almost identical and thus the lower resolution mesh was chosen to reduce the computational costs during the forward calculation.

5.4.3 Elastic 3D Image Registration

5.4.3.1 Project Requirements

Unlike in previous studies [258, 260], the contraction of the ventricles that was investigated here was not computed with a biomechanical model but rather derived

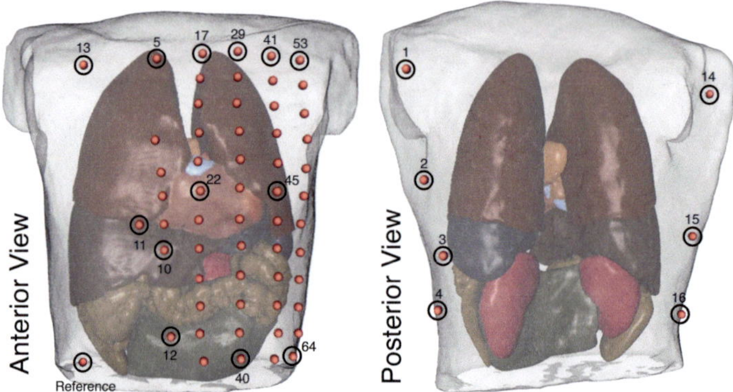

Fig. 5.10. Voxelized model of the torso containing the following organs: lungs, liver, spleen, kidneys, small intestine, colon, skeletal muscle (not visible), anisotropic heart muscle, blood and fat tissue. Electrode positions at which body surface potentials were extracted are marked by red spheres. 57 electrodes were positioned on the torso front, while 7 electrodes were situated on the back.

from MR image data (Cine and tagged MRI). However, in order to be able to consider the ventricular deformation during the forward calculation, the trajectories of the cardiac nodes and thus the movements of the associated electrical sources have to be known. The biomechanical models introduced earlier [258, 260] directly provide this information as the trajectories are calculated and thus readily available. Yet, in this study, the ventricular deformation is described by 30 different anatomical models that were created based on Cine MRI data. A drawback of these Cine images is that they only contain information on the movement of the endocardial and epicardial surfaces. Besides the coarsely spaced slices with tagging data, no information was available that described the movement of the cardiac nodes within the ventricular walls.

To obtain the trajectories of nodes inside the walls, we chose to use an elastic registration technique. During each registration, the diastolic dataset of the heart and surrounding organs was matched onto the anatomical models in the contracted states that were described previously (see 5.4.2.2). The result of each registration process was a deformation field that characterized the movement of each individual node and could be used to model the ventricular deformation. In total, 29 registration procedures were carried out in order to construct the dynamic model. Based on the assumption that the movement of the ventricles should only affect tissues in direct vicinity, we introduced a bounding box around the heart and considered only the movement of tissues within this box during the registration. More details on the bounding box and technical implementation will be provided in section 5.4.3.3.

5.4.3.2 General Considerations of Image Registration

In general, elastic image registration is an ill-posed problem and thus a direct solution approach is not possible [329]. Other approaches that are often used and suitable for these kind of registration problems are based on a similarity measure and a regularizer. The similarity measure can be considered as the driving force whereas the regularizer controls the deformation. In case of this study, the registration should match differently shaped ventricles onto each other. Therefore a regularizer that considers the physical properties of the heart and surrounding tissues would be ideal.

Recently, registration methods for medical images have been proposed that use key features of biomechanical models [330, 331]. In these methods, the imaged objects are modeled as elastic bodies that are deformed by image similarity forces while at the same time, the biomechanical elasticity model is used to regularize this deformation.

One possibility to calculate image similarity forces is based on the so-called Iterative Closest Point (ICP) algorithm. In the past it has mainly been used to estimate the *rigid* transformation of roughly aligned 3D datasets. Since its first introduction [332, 333], the algorithm has been used e.g. in the integration of range images [334], for the alignment of MRI images [335] and human motion tracking [336]. When initialized with two 3D point clouds, a source cloud S and a target cloud T, the algorithm tries to iteratively find a rigid transformation that minimizes a similarity metric of all corresponding pairs of source and target points. In each iteration k, the algorithm performs two main tasks:

- Define correspondence pairs between points in S and T
- Find the rigid transformation that minimizes an error metric that depends on the previously determined correspondence pairs (e.g. mean square error of distances between source and target point)

Each correspondence pair consists of a point $\mathbf{x}_{k,i}(i = 1,\ldots,m)$ from S and the point $\mathbf{x}_{k,j}(j = 1,\ldots,n)$ from T that satisfies a correspondence metric. Often the minimal Euclidean distance to $\mathbf{x}_{k,i}$ denoted $d_{k,i}$ is used:

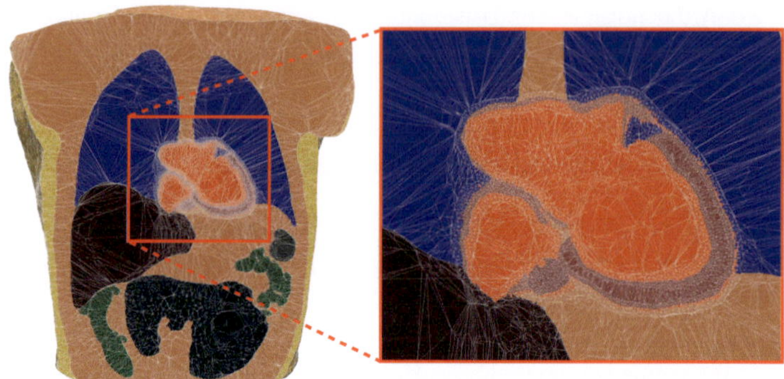

Fig. 5.11. Sliced view of the unstructured tetrahedron mesh of the torso that was used for the forward calculation of the body surface potentials. The node density inside the heart was higher than in the rest of the torso. The red box marks the bounding box that was used during the registration procedures. The movement of all nodes within this box was considered during the registration. The nodes on the surface of the box and all nodes outside the box were fixed and thus prevented from moving.

$$d_{k,i} = \min_{j \in \{1,\cdots,n\}} \|\mathbf{x}_{k,j} - \mathbf{x}_{k,i}\| \tag{5.1}$$

Since its initial introduction, many variants of the ICP algorithm have been proposed [337]. Recently, Amberg *et al.* showed how to extend the ICP algorithm to the *nonrigid* registration of surfaces while retaining the convergence properties of the original algorithm [338]. In order to impose constraints on the deformation, a regularizer was used that minimized the difference between transformations acting on neighboring mesh vertices. This can be interpreted as a kind of stiffness term that forces neighboring vertices to undergo similar transformations [338]. The approach that was used in this study to register the surface of the diastolic ventricles to the surface of the deformed states used a similar regularization principle as Amberg *et al.* [338]. Details will be provided in the next section.

5.4.3.3 Current Implementation

As explained previously, we assumed that the ventricular contraction only affected tissues that were in direct vicinity of the heart and thus lay inside a pre-defined bounding box. While the ventricles contract and relax, the adjoining tissues follow the deformation passively. To consider this passive movement while matching the ventricular surfaces onto each other, a combination of the nonrigid ICP algorithm and a biomechanical model of elasticity was used [339]. The elastic image registration procedure described below was conducted in close cooperation with Oussama Jarrousse and Thomas Fritz.

The bounding box (180 x 148 x 136 mm^3) contained the heart in diastolic state and all surrounding tissues that consisted out of fat, muscle, lung, liver, spleen and colon. Vertices located on the surface of the bounding box were fixed and thereby prevented from moving. A visualization of the bounding box is shown in Fig. 5.11. Details on the elasto-mechanical parameterization of the tissues inside the bounding box can be found in [13].

To improve the quality of the registration, four subsets of source and target points were defined. Points from one of the source subsets will only search for correspondences in the associated target subset. The subsets contained the endocardial and epicardial surface of the left and right ventricle that were extracted from the models created based on the Cine MRI data.

In addition to this surface information, tagging data that described the movement within the ventricular wall was included into the dynamic model as well. 23 tagging points (18 from the left ventricle and 5 from the right ventricle) were manually tracked in three slices over the whole cardiac cycle. As tagged and Cine MR images had different temporal resolutions, the movement of the landmarks was linearly interpolated so that the respective time bases matched. An example of the short axis tagging data can be seen in Fig. 5.12 at selected phases of ventricular contraction and relaxation. Manual extraction of the tagging information was difficult as the contrast of the tagging grid was not constant over time and therefore hard to track.

Each of the 29 iterative registration procedures adhered to the following scheme: In each iteration, correspondences were defined and a cost function that depended on the vertice coordinates was minimized. The correspondences were defined based on the minimal Euclidean distance (see equation 5.1).

To finally determine the deformation, a cost function $\mathbf{E}$ was minimized (similar as in [338]). Here, the cost function was given by:

$$\mathbf{E}(\mathbf{X}_k, k) = \mathbf{E}_\varepsilon(\mathbf{X}_k) + \mathbf{E}_{icp}(\mathbf{X}_k, k) + \mathbf{E}_\lambda(\mathbf{X}_k, k) \qquad (5.2)$$

Where $\mathbf{X}_k$ is a vector of all vertice coordinates of the model in iteration k. $\mathbf{E}_\varepsilon(\mathbf{X}_k)$ is related to the biomechanical model of elasticity and consists of a set of strain energy density functions governing the different passive elastic properties of the model. It can be interpreted as the stiffness of the model.

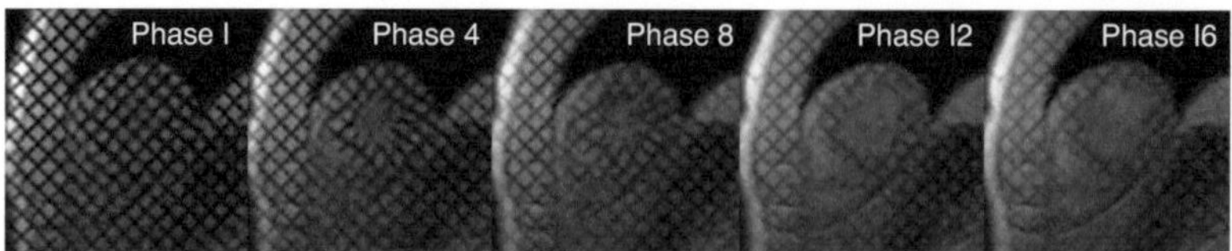

Fig. 5.12. This figure shows one of the three tagging slices that was used to extract landmark information for the registration procedures. The tagging data had a lower temporal resolution than the Cine data (tagging data: 18-22 phases vs. Cine data: 50 phases). Linear interpolation was used to calculate the locations of the landmarks in between the imaged phases. Furthermore, the non-uniform contrast can be seen when the different phases of the tagging data are compared (e.g. phase 1 with phase 16). This additionally impeded the manual extraction of the landmarks.

$\mathbf{E}_{icp}(\mathbf{X}_k,k)$ and $\mathbf{E}_\lambda(\mathbf{X}_k,k)$ are determining the image similarity forces. $\mathbf{E}_{icp}(\mathbf{X}_k,k)$ is given by:

$$\mathbf{E}_{icp}(\mathbf{X}_k,k) = k\sum_{i=1}^{m}\alpha_s d_{k,i}^2 \tag{5.3}$$

where m is the number of source points, α_s is a weighting parameter for the endocardial and epicardial surface points and $d_{k,i}$ is the Euclidean distance between the correspondence pair $(\mathbf{x}_{k,i}, \mathbf{x}_{k,j})$ as presented earlier in equation (5.1).

Similarly the landmarks term $\mathbf{E}_\lambda(\mathbf{X}_k,k)$ is given by:

$$\mathbf{E}_\lambda(\mathbf{X}_k,k) = k\sum_{l=1}^{p}\alpha_l d_l^2 \tag{5.4}$$

where p is the number of landmarks, α_l is a weighting parameter specific for the landmarks and d_l is the Euclidean distance between source and corresponding target landmark.

A suitable solution for the registration problem was found when the cost function of equation 5.2 was minimized. In this case, the criteria of both, the image similarity measures and the biomechanical elasticity model, were satisfied. The associated non-linear system was solved in parallel with an implementation of the Newton method from the PETSc package. After each iteration, the vertices' positions were updated and the registration continued until a termination criterion was reached. In this project, the termination criterion was reached when both, the mean distance $\bar{d}$ of all correspondence pairs and landmarks was below 0.07 mm, and the maximum distance d_{max} was below 0.5 mm. This was a good compromise between computational costs and fidelity of the registration results.

From equation (5.2), (5.3) and (5.4) one can conclude that at the beginning of the registration process (small k) the model's stiffness E_ε (independent of k) dominates the deformation. However, with increasing iterations (growing k), the image similarity terms (E_{icp} and E_λ, both depending on k) become larger and finally overcome the stiffness of the biomechanical model.

The result of each registration procedure was a deformation field. It was used to move the nodes of the unstructured tetrahedron mesh of the torso that was used for the forward calculation to their position in the dynamic model. This was done by iterating through the nodes inside the bounding box (see inlet Fig. 5.11) and calculating the new position by trilinear interpolation from the deformation field. The fiber orientation was adapted to the contracted states in a similar way. In this work, the fiber orientation of each tetrahedron was defined by the vector connecting the center of gravity and a second point lying in the direction of the fiber orientation. To determine the fiber orientation in the contracted state, both points were moved based on trilinear interpolation of the deformation field. The final position of these two points was used to calculate the fiber orientation in the deformed state. The adapted unstructured tetrahedron meshes and fiber orientation were subsequently utilized for the dynamic forward calculations. Deformed states in between the 29 registered frames were interpolated linearly such that the resulting dynamic torso model had a temporal resolution of 2 ms (i.e. every 2 ms the deformation state of the model was updated).

In total, three different anatomical torso models were created for the forward calculation. They contained the ventricles in dynamic and static conditions, respectively. The following setups were investigated as to their influence on the T-Wave in the body surface ECG:

- STATIC denotes a static torso model in which the diastolic state of the ventricles was used for the whole cardiac cycle.
- DYNA denotes a dynamic torso model. While considering ventricular contraction and relaxation, this model discards the deformation information provided by the manually tracked landmarks (i.e. $\mathbf{E}_\lambda(\mathbf{X}_k, k) = 0$, see (5.2)).
- DYNALAND denotes a dynamic torso model as well. In contrast to the DYNA model, this model includes the displacement information that was derived from the landmarks. In this case, α_l was set to $10 \cdot \alpha_s$ to emphasize the influence of the landmarks.

5.4.4 Electrophysiological Modeling

5.4.4.1 Endocardial Stimulation Profile

For the realistic simulation of the body surface ECG, it is imperative to model a realistic depolarization sequence. In this case, ventricular activation was determined by a specialized endocardial stimulation profile that mimicked the role of the excitation conduction system. The stimulation profile was created by adapting a semi-automatic approach from Werner *et al.* [185, 2]. Details on the construction of the endocardial stimulation profile can be found in [2, 13] and in section 4.1. The parameters of the stimulation profile were manually adapted in order to achieve a good match between the simulated ECG and a clinical recording. The resulting stimulation profile is visualized together with the corresponding isochrone maps in Fig. 5.13A and 5.13B.

5.4.4.2 Electrophysiological Heterogeneities

In this study, the first version of the ten Tusscher model for human ventricular myocytes was used to describe the electrophysiological properties of the ventricular cells. The associated mathematical equations were solved with a combination of Rush-Larsen and forward Euler method (for details, see [13]).

To allow for a realistic repolarization sequence, transmural and apico-basal heterogeneities were included into the model. For a list of experimentally measured heterogeneities and their influences on the T-Wave morphology please refer to section 4.2. In this case, heterogeneous properties of the slowly delayed rectifier current I_{Ks} and the transient outward current I_{to} were considered. Endocardial, M and epicardial cells were assumed to appear in layers. Layer thickness was 40% for endocardial, 40% for M and 20% for epicardial cells. This is a good compromise between the more endocardial position of the M cells reported in [155, 262, 272] and the more epicardial position that is described in [261].

According to [262, 261, 340, 341, 155] tissue resistivity is not distributed uniformly across the ventricular wall. A sudden drop in gap junction density between M and epicardial cells has been reported to reduce electrotonic coupling allowing large APD_{90} gradients at the respective locations. To consider these effects in the model, resistivity scaling factors have been extracted from the wedge-measurements by Yan *et al.* and were used to adapt transmural conductivites in the left and right ventricle accordingly.

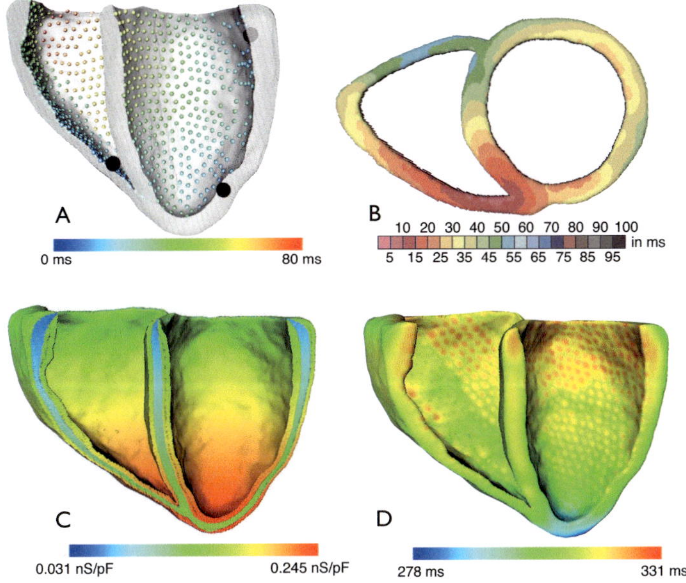

Fig. 5.13. A: Color-coded stimulation profile together with the root-points (black spheres) that were used during the construction of the stimulation profile as described in [2, 13]. B: Isochrone map of the activation sequence in a horizontal plane at mid-to-apical height. Color scale and cut-plane are adopted from the work of Durrer *et al.* [174] to facilitate the comparison with their invasively acquired isochrone measurements. C: Heterogeneous g_{Ks} distribution. The layered distribution of endocardial, M and epicardial cells is clearly visible. In apico-basal direction, a linearly increasing density of g_{Ks} was implemented. D: APD_{90} distribution for the physiological setup with normal QT time as explained in section 5.4.5.

In apico-basal direction, a higher apical g_{Ks} was assumed based on [263]. In measurements on dogs, apical g_{Ks} was found to be approximately twice as large compared to basal density [263]. A similar distribution although with a weaker gradient was also found in human left ventricular tissue [263]. g_{Ks} distribution in the model was implemented accordingly ($g_{Ks,Apex} = 2 \cdot g_{Ks,Base}$). Values in between apex and base were linearly interpolated. The distribution of g_{Ks} together with the resulting APD_{90} for a setup with physiological QT times (see 5.4.5) can be seen in Fig. 5.13C and 5.13D.

5.4.5 Modeling physiological QT times, SQT and LQT

In order to guarantee a realistic relation between mechanical relaxation and electrical repolarization for physiological QT times, we had to ensure that the QT time of the simulation was comparable to the proband's QT time during the MR image acquisition. As the QT time was not directly extractable from the MRI system, we determined the heart rate during the MRI scan and used it to correct the clinically acquired ECG. Bazett-corrected QT time was approximately 391 ms (manually extracted from Einthoven II), which is in good agreement with values from Mason *et al.* for the corresponding age group [134]. In case of the simulation setup for physiological QT times, no parameter adaptions had to be made as the standard parameterization initially proposed by ten Tusscher *et al.* delivered a similar QT time.

The ion channelopathy termed SQT syndrome has been associated with gain-of-function mutations in KCNH2 (HERG, SQT1), KCNQ1 (KvLQT1, SQT2), and KCNJ2 (Kir2.1, SQT3). Patients suffering from this disease have shortened QT interval durations ($QT_c < 330$ ms) and tall peaked T-Waves. The SQT syndrome was modeled by simply increasing the maximum conductivity $g_{Kr,max}$ of the slowly delayed rectifier current I_{Kr} by a factor of 2.5 to shorten the QT interval.

The LQT syndrome can be caused by mutations in various different ion channels [103]. Affected patients show QT prolongation ($QT_c > 450$ ms) and changes in T-Wave morphology depending on the LQT subtype [103]. For more details on the LQT syndrome, please refer to section 2.2.5 and section 3.1.6. Here, LQT was modeled by a complete block of I_{Kr} ($g_{Kr,max} = 0$).

The action potential propagation in the tissue was described based on the monodomain reaction-diffusion model. The associated mathematical equations and implementation details are described in section 3.1.3.

5.4.6 Forward Calculation of the ECG in the Static and Dynamic Models

Input data for the forward calculations were the transmembrane voltage distributions that resulted from the simulations of depolarization and repolarization on the static diastolic voxel model of the left and right ventricle. These distributions were interpolated onto the static unstructured tetrahedron mesh (see Fig. 5.11).

In case of the STATIC model, the transmembrane voltages assigned to the nodes in the static unstructured tetrahedron mesh were used to calculate the ECG over the whole cardiac cycle. In case of the dynamic torso models (DYNA / DYNALAND), the transmembrane voltage distributions for each time step were interpolated onto the static tetrahedron mesh similarly as for the STATIC model. Yet, after the interpolation, the nodes were moved to their corresponding positions in the dynamic torso models before the forward calculation was performed. To describe the current deformation state in the dynamic models, a new tetrahedron mesh was used every 2 ms.

During the forward calculations, the bidomain equations were utilized to calculate the extracellular potentials. Details on the solution of the forward problem can be found in [15] and in section 3.2.3. The resulting body surface potentials were extracted at the 64 electrodes that covered the front and parts of the back of the proband's torso (electrode position is shown in Fig. 5.10).

5.5 Ranking the Influence of Different Tissues with Respect to their Conductivities

As explained in section 3.2.3.2, tissue conductivities can have an important effect on the results of the forward problem of electrocardiography. Yet, although there is a large uncertainty in reported conductivity values for different organs, no study has so far addressed this issue and the potential ramifications in a systematic way. In this section we present a study that has the following aims:

- A sensitivity evaluation was conducted to evaluate the most important inhomogeneities for a realistic solution of the forward problem. To this end, we used fixed percental conductivity changes and ranked the inhomogeneities according to their importance.
- An uncertainty analysis evaluated the effects of contradictory conductivity reports between different experimental studies. In this case, the tissue conductivities were set to the reported minimum and maximum values and the resulting BSPM changes were considered in the subsequent ranking.
- Finally we removed groups of organs from our torso model and calculated the associated errors with respect to a reference model. This was done to be able to propose recommendations that are targeted on facilitating the creation of patient-specific models.

Unlike some existing studies with similar (but not identical) scopes, the solution of our forward problem was based on a distribution of cardiac sources that resulted from the realistic simulation of a complete heart beat. Therefore the presented results are particularly applicable to future studies that aim at clinical applications of cardiac simulations.

The results of this study can be found in section 7.5.

5.5.1 Electrophysiological Modeling

The Visible Man dataset [309] was chosen to investigate the impact of the different tissue conductivities on forward calculated ECGs. Due to the fusion of information from different imaging modalities (MR, CT, thin section photography) the Visible Man dataset provides highly detailed anatomical information that is unmatched by models that are created based on clinically available image data.

Cardiac depolarization and repolarization (in sinus rhythm) were simulated separately for atria and ventricles and was used as input data for the calculation of the body surface potentials. Different models to simulate the action potential propagation in the tissue were used for atria and ventricles. For the atria, a rule-based Cellular Automaton was used to calculate the distribution of transmembrane voltages and the spread of excitation [185]. In contrast to that, the current flow in the ventricles was described using the monodomain reaction-diffusion model in conjunction with a biophysically detailed ionic model [151]. This allowed the inclusion of complex electrophysiological heterogeneities in the ventricles, which are important for a realistic repolarization sequence.

Atrial excitation was simulated in a voxelized model with an isotropic resolution of 0.33 mm. The electrophysiological model from Courtemanche, Ramirez and Nattel [342] was used to parameterize the Cellular Automaton. According to reports from Hansson *et al.* the conduction velocity was set to 700 mm/s [343]. Faster conductivities were assigned to the crista terminalis (1300 mm/s) and Bachmann's bundle (1770 mm/s) [344, 345].

For the simulation of ventricular excitation, the anatomical dataset was interpolated to an isotropic voxel size of 0.4 mm. Anisotropy of ventricular conduction was considered by incorporating muscle fiber orientation as described in section 3.2.2 and [15]. The electrophysiological properties of the ventricular tissue were described using the model developed by ten Tusscher *et al.* [151]. The model contains transmurally heterogeneous descriptions of the ion channel characteristics of the slow delayed rectifier current I_{Ks} and the transient outward current I_{to}. In addition to these transmural heterogeneities we modeled apico-basal gradients of g_{Ks} to enable a realistic repolarization sequence. Details on the modeling of the included electrophysiological heterogeneities can be found in [15] and in section 5.4.4. Ventricular activation was initiated by a sequence of endocardially applied stimuli currents. The model that determined the location and temporal sequence of these stimuli currents was described in [2] and in section 4.1.

Before simulating the ventricular activation, all heterogeneous parameter configurations of the electrophysiological model were pre-calculated in a single cell environment (duration of the pre-calculation: 60 s with a basic cycle length of 1 s). This was done to tune gating variables and ionic concentrations. Furthermore, the intracellular conductivities were adapted so that they compensated for the large voxel size of 0.4 mm and the thicker walls of the Visible Man dataset. The adapted intracellular conductivity was chosen such that the average transmural conduction time was approximately 30 ms [262].

5.5.2 Torso Model, Conductivities and Forward Calculation

In order to solve the forward problem of electrocardiography, the voxelized torso model was converted into an unstructured tetrahedron mesh. The following tissues and structures were considered in the model: heart, blood (both intracavitary and in the main vessels), lungs, fat (both visceral and subcutaneous), anisotropic skeletal muscle (will be referred to as muscle in all tables), intestine, liver, kidneys, bone, cartilage, and spleen. Tissues covering less than 0.5 % of the body volume are omitted in this list.

The construction of the tetrahedron torso geometry was initialized by creating a mesh with 70,000 nodes that characterized the contours of the torso and the shape of the internal organs. The nodes of this mesh were chosen from a 2 mm voxel dataset of the torso. Subsequently the heart region was refined by adding 200,000 additional nodes based on the high-resolution cardiac datasets described in section 5.5.1: 190,000 for the atria and ventricles and 10,000 for the major blood vessels around the heart.

In the tetrahedron torso model, the anisotropic electrical conductivities of the ventricles and skeletal muscle were considered. Ventricular fiber orientation was adopted from the simulations of ventricular excitation described in section 5.5.1. In case of the skeletal muscles, fiber orientation was extracted from the highly detailed thin-section photos of the Visible Man dataset [253]. Automatic methods such as texture analysis and a 3D Sobel filter were used to derive an initial orientation estimation, which was then revised by human experts. The resulting skeletal muscle fiber setup is shown in Fig.5.14.

In our standard setup, we used the tissue conductivities that were reported by Gabriel *et al.* at 10 Hz [245]. Skeletal muscle anisotropy ratio was set to 7 as this seems to be the value that is most frequently cited in the literature [241]. In

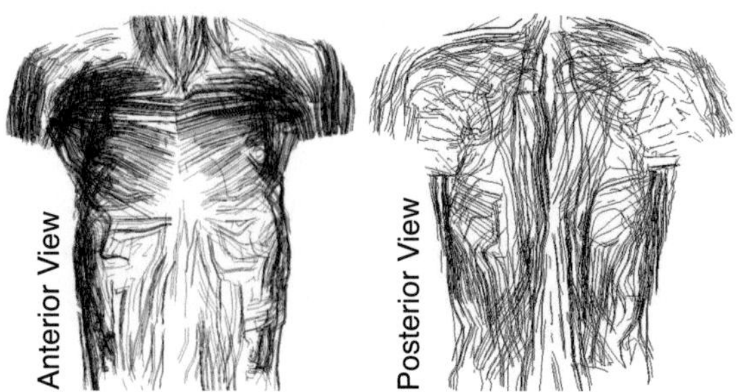

Fig. 5.14. Skeletal muscle fiber orientation of the Visible man dataset [253]. The orientation information was extracted from the highly detailed thin-section photos using a combination of automatic methods such as texture analysis, a 3D Sobel filter and manual corrections.

the Visible Man model, the intestine was not separated into small intestine and colon. Therefore we used an average intestine conductivity of 0.278 S/m, which was calculated based on medical textbook's length and diameter statements on the small intestine and colon (for details see [15]) and the corresponding conductivities reported by Gabriel *et al.* at 10 Hz [245]. The resulting volume fractions of the small intestine and colon were 53.3% and 46.7% respectively.

As explained in section 3.2.3.2 there are a number of reasons for differences in conductivity measurements originating from different studies. In order to evaluate these conductivity uncertainties, a table was compiled listing the upper and lower boundary of reported conductivities for the respective organ. Ten primary sources ([247, 249, 250, 346, 347, 348, 349, 350, 351, 352]) and five review articles ([241, 246, 245, 248, 353]) were considered. If a range of conductivities instead of a single value was reported from a measurement paper, we considered the respective upper or lower boundary for the listing. An exception was made for the intestine: as only one literature source provided measurement data on the small intestine and colon and due to the fact that most anatomical torso models do not separate between the different parts of the intestine, we chose the lower boundary of the intestine conductivity based on measurements of colon samples and the upper boundary based on the reported value for the small intestine. As the forward problem is considered to be a quasi-static problem, all measurements that were performed at frequencies above 10 kHz were excluded from the listing.

Due to the scarcity of measurement data from human samples, most simulation studies also use conductivity values from animal studies. Therefore we also in-

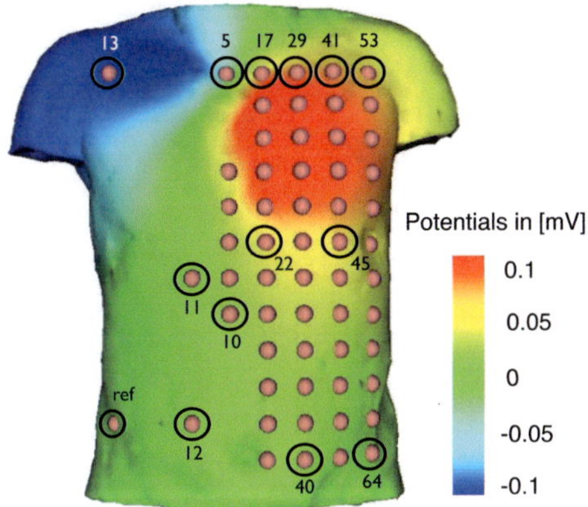

Fig. 5.15. Unstructured tetrahedron model of the Visible Man torso with a superimposed BSPM showing the atrial activity at 41 ms after sinus node stimulation. Visualized electrodes mark the locations were the body surface potentials were extracted for subsequent analysis. Seven electrodes were located on the back of the torso and are therefore not visible.

cluded measurements on animal samples into our listing. Finally no differentiations were made between in-vivo or ex-vivo measurements or investigations at different sample temperatures.

The forward calculations were performed based on the classic two step approach: As a first step, atrial and ventricular activation and repolarization were simulated on a voxelized model of the heart as described in section 5.5.1. Then the resulting transmembrane voltages were interpolated onto the unstructured tetrahedron mesh of the heart and the bidomain model [171] was used to calculate the extracellular potentials in the whole torso. Details on the forward problem and the associated mathematical equations can be found in [15] and in section 3.2.3.

The resulting body surface potentials were extracted at 64 electrodes that cover the torso front and parts of the back of the Visible Man model. The electrodes' positions can be seen in Fig. 5.15. The signals that were extracted at each electrode were rearranged into an n-dimensional spatio-temporal vector Φ for further analysis. Assuming there are m electrodes and each electrode records the body surface potentials at t samples, the vector dimension n can be calculated by $n = m \cdot t$.

5.5.3 Ranking the Conductivities with respect to Sensitivity and Uncertainty

To evaluate the importance of a certain tissue or fluid within the torso model, we used two different ranking methods:

1) In a *sensitivity analysis*, we probed changes of the body surface potentials by increasing and decreasing the conductivity of one organ at a time by 25% of the values reported by Gabriel *et al.* [245]. A similar evaluation was done for the skeletal muscle anisotropy which was set to 5.25 and 8.75, respectively. The difference between the BSPMs that were calculated using the increased and decreased conductivity was a measure for the influence of the associated organ.

2) In an *uncertainty analysis*, the conductivity of each organ was set to the minimal and maximal conductivity that was reported in the literature while the conductivities of all other organs remained at their standard values. This was a measure that could be used to evaluate the influence that the currently existing measurement uncertainties can have on the simulated body surface potentials. An example of the potential changes that can be associated with measurement uncertainties is shown in Fig. 5.16.

We used three different quantitative measures for both the sensitivity and the uncertainty ranking. I.e. we used the root mean square error (RMSE) to evaluate

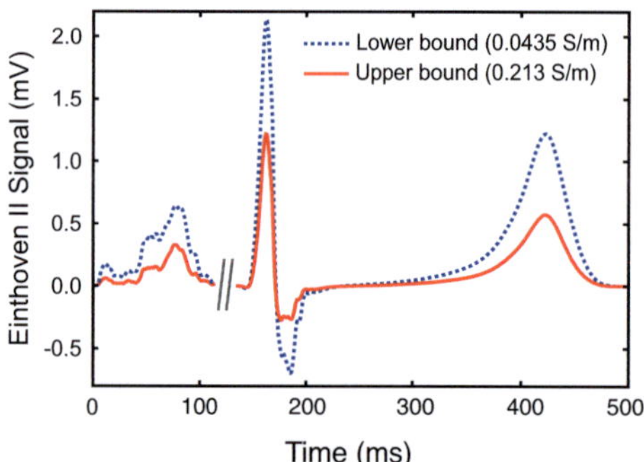

Fig. 5.16. ECG in the Einthoven II lead resulting from forward calculations based on different skeletal muscle conductivities. Atrial and ventricular activation was unchanged between both visualized signals. The skeletal muscle anisotropy ratio was fixed to 7. The signals correspond to the lowest (0.0435 S/m [347]) and highest (0.213 S/m [346]) skeletal muscle conductivity obtained from the literature. The Einthoven II lead was calculated using electrode 13 and 64 (see Fig. 5.15).

differences in signal amplitudes between the signals Φ_1 and Φ_2 calculated at two conductivity values σ_1 and σ_2

$$\text{RMSE} = \sqrt{\frac{1}{n} \cdot \sum_{i=1}^{n} [\Phi_1(i) - \Phi_2(i)]^2} \qquad (5.5)$$

A second measure that was not influenced by homogeneous signal scaling but rather considered changes in signal morphology or changes between different signal regions or time instants was the normalized RMSE ($\text{RMSE}_{\text{norm}}$). It was calculated by normalizing each signal vector to a maximum absolute value of 1 before calculating the RMSE.

Finally, the linear correlation coefficient (CC) was used as third measure. Between the signals Φ_1 and Φ_2 it can be calculated by:

$$\text{CC} = \frac{1}{s_1 \cdot s_2} \sum_{i=1}^{n} \left[\Phi_1(i) - \overline{\Phi}_1\right] \cdot \left[\Phi_2(i) - \overline{\Phi}_2\right] \qquad (5.6)$$

Here, $\overline{\Phi}_{1,2}$ are the arithmetic mean values and $s_{1,2}$ the standard deviations of the respective signals $\Phi_{1,2}$.

5.5.4 Possible Torso Model Simplifications

In order to assess possible torso model simplifications, we removed a varying number of low-ranking organs from the model. If the associated body surface potentials were similar to the results that were calculated with a fully heterogeneous torso model, the removal of these organs during the creation of future torso models would be permitted. Every simplified setup contained the heart as the cardiac anatomy is a prerequisite for the simulations of the transmembrane voltages that are used as input for the forward calculations. In addition to the heart, the 7, 5 or 3 most important organs or structures of the RMSE/CC-sorted atrial and ventricular sensitivity ranking were added (see Table 7.5). The resulting setups were named $\text{TOP7}_{\text{RMSE}}/\text{TOP7}_{1-\text{CC}}$, $\text{TOP5}_{\text{RMSE}}/\text{TOP5}_{1-\text{CC}}$ and $\text{TOP3}_{\text{RMSE}}/\text{TOP3}_{1-\text{CC}}$. As some research groups also use homogeneous torso models we evaluated the quality of the associated results by creating a homogeneous model. In this case, HOM_{RMSE} and $\text{HOM}_{1-\text{CC}}$ were identical.

The virtual removal of organs was performed by replacing the respective organs with a mean conductivity $\overline{\sigma}$ that represented the averaged conductivity within the torso. Two different mean conductivity values were considered:

1) $\overline{\sigma}_1$ was based on the volume fraction of the different organs and the conductivity values from Gabriel *et al.* at 10 Hz [245] that were associated with the respective organ. This weighted average conductivity was calculated to be 0.123 S/m.

2) $\overline{\sigma}_2$ was based on literature values for the average torso conductivity. Two studies were available reporting average torso conductivities of 0.241 S/m [348] and 0.216 S/m [350]. We used the arithmetic mean of these two reported values: 0.229 S/m. As mentioned above, the performance of the simplified setups was rated by comparing the associated results with the fully heterogeneous torso model which served as gold standard. Evaluation criteria were the RMSE or CC (depending on which criterion was chosen for the selection of the most important organs).

5.6 BSPM Prediction for Varying Conductivities Based on PCA

In section 7.5 we will show how the BSPM can be influenced by changes in tissue conductivities and we will recommend a list of organs that should be included in a torso model which is used for the solution of the forward problem. However, also within these most important organs the tissue conductivity is not precisely known (reasons for this conductivity uncertainties were introduced in section 3.2.3.2). Presently, BSPM changes that are due to conductivity variations can only be assessed by repetitive forward calculations. A disadvantage of such an approach is that a large number of forward calculations would have to be performed, which is very time consuming and probably not possible in a potential future clinical application.

In this study, we therefore propose a more efficient approach based on the Principal Component Analysis (PCA) to overcome these limitations. The PCA is a statistical method that allows to describe the variance in data by transforming it to a new set of orthogonal basis vectors. These new basis vectors are chosen such, that the representation error is minimized if the dimensionality of the data is reduced [354]. The PCA has been used in many application areas of biomedical engineering, e.g. in image processing [355]. When it comes to ECG or BSPM analysis, it has been utilized to remove spatial redundancy [356], extract respiratory information [357], estimate T-Wave alternans [358], or suppress signal noise [359, 360].

In this study, we use the PCA to predict conductivity related BSPM changes from few sample simulations. To this end, we performed seven forward calculation for each considered organ (blood, muscle, lung, fat). In each forward calculation, the conductivity was varied between $\pm 75\%$ of the default value (steps of 25%). The resulting signals were then fed into the PCA and allowed to estimate the ECG over the whole sampled conductivity range. This was possible as conductivity induced BSPM variations were described by a mean signal and the first PCA eigenvector scaled by a conductivity dependent PCA score. Due to the monotonic nature of the PCA score curve, missing scores could be interpolated and we were thus able to reconstruct the BSPM for conductivities that were not part of the initial sample.

We evaluated this technique for conductivity changes in a single tissue as well as for changes in multiple tissues at the same time. In addition to that, the proposed method can be used to calculate confidence intervals for a simulated ECG, which constitute the upper and lower signal boundaries for arbitrary conductivity uncertainties. Finally we used the PCA method in the opposite direction: rather than determining the most likely BSPM signal for a set of known conductivities, we probed hundreds of different conductivity combinations in an effort to find the most likely conductivities for a given BSPM based on a numerical optimization scheme.

The results of this study can be found in section 7.6.

5.6.1 Electrophysiological Modeling

The highly detailed Visible Man dataset [309] provided the underlying anatomy for the simulation of cardiac excitation and repolarization and the subsequent solution of the forward problem of electrocardiography. Action potential propagation in the atria and ventricles was described with the monodomain-reaction-diffusion model (see section 3.1.3) that was integrated into a C++ simulation framework [325].

For the simulation of atrial excitation in normal sinus rhythm, the Visible Man atrial dataset was interpolated to an isotropic voxel size of 0.3 mm. Atrial electrophysiology was described based on the model by Courtemanche *et al.* [342], which was initialized with 60 beats at 1 Hz. Intracellular conductivities were chosen such that the conduction velocity in the isotropic atrial tissue was 70 cm/s. In the simulations, atrial activation (P-Wave) was completed after 150 ms.

Ventricular excitation and repolarization were simulated in an anisotropic model of the Visible Man dataset that accounted for transmural and apico-basal electrophysiological heterogeneities. Details concerning the fiber orientation, the electrophysiological model and its heterogeneous parameterization can be found in [18, 15]. Ventricular activation was initiated by a special sequence of endocardial stimulations that mimicked the role of the excitation conduction system, see section 4.1 and [2]. Ventricular activation and repolarization was simulated for a total duration of 400 ms (comprising QRS complex and T-Wave).

5.6.2 Torso Model and Forward Calculation

The unstructured tetrahedron model of the Visible Man dataset that was used for the solution of the forward problem was adopted from [15] and is described in detail there. It contained the following tissues: anisotropic heart muscle, blood (both intracavitary and in the main vessels), lungs, fat (both visceral and subcutaneous), anisotropic skeletal muscle (from now on referred to as muscle), intestine, liver, kidneys, bone, cartilage, and spleen. For details on the modeling of ventricular or skeletal muscle anisotropy please refer to [15]. Anisotropy ratios (along:across conductivity) were set to 3:1 for the ventricular tissue and 7:1 for the skeletal muscles. The conductivities used in this study were based on the values published by Gabriel *et al.* at 10 Hz [245] (from now on referred to as GG). The only exception was made for the conductivity of the intestine for which we used an averaged conductivity of 0.278 S/m as calculated in [15].

The forward calculations were again performed using the common two-step approach. At first, cardiac excitation was simulated as described in section 5.6.1. Then the resulting distribution of transmembrane voltages was interpolated on the unstructured tetrahedron model of the torso and the bidomain model [171] was used to calculate the extracellular potentials (for details see section 3.2.3 and [15]). For the subsequent evaluation, the body surface potentials were extracted at 64 electrodes which were positioned as shown in 5.15. The electrodes mainly covered the central and left side of the thorax where highest signal variability is expected. The following organs, tissues and fluids (from now on simply referred to as tissues) are known to have a strong influence on the body surface ECG: blood, muscle, lungs, fat [15, 240, 243] (see section 7.5.2). To probe our PCA-based BSPM prediction technique we chose to evaluate variations in these most important tissues. To this end, we conducted forward calculations with varying conductivities for each organ and fed the results as input data into a PCA analysis. For each of the four tissues, we performed seven forward calculations at the GG conductivity and GG$\pm$25%, GG$\pm$50% and GG$\pm$75% to account for the existing conductivity uncertainties due to measurement difficulties, sample variations or conductivity affecting diseases. A change of $\pm$75% translates into a ratio of 7 between upper and lower conductivity boundary. Typical uncertainty ranges for the four tissues under investigation are between 2.3 and 5.5 (see [15] and Fig. 7.16), so a maximal change of $\pm 75\%$ should be sufficient to cover this span. To enhance readability, the seven conductivities will be referred to as σ_i = -75%, -50%, -25%, GG, +25%, +50%, +75% in the following.

5.6.3 BSPM Analysis using PCA

As explained earlier, changes in tissue conductivities lead to changes in the associated BSPMs. The changes were quantitatively analyzed based on the PCA as will be explained in the following. We separated our analysis between atrial and ventricular signals. As the PCA decomposition was applied separately as well for each tissue (blood, muscle, lungs, and fat) and for atrial and ventricular signals, eight PCA decompositions had to be performed in total.

5.6.3.1 Assembly of the Spatio-Temporal Data Matrix

PCA is a statistical analysis method to detect patterns in data of high dimension. Usually the data which should be analyzed is stored in a matrix $\mathbf{X}$ that consist of m variables with n observations each and thus has $m \times n$ entries. In this study, the n observations corresponded to the different conductivities that were under investigation (i.e. $n = 7$) while each conductivity delivered a signal consisting of m data points (m = number of electrodes $\times$ number of time steps).

In other words, the forward calculations resulted in seven BSPMs from the different conductivity values, recorded at 64 electrodes at time steps $t = t_0, \ldots, t_{\max}$. For every conductivity σ_i, the signals from all 64 electrodes at all time steps could therefore be concatenated into one spatio-temporal signal vector

$$\mathbf{x}_{\sigma_i} = [x^1_{\sigma_i}(t_0), \ldots, x^1_{\sigma_i}(t_{\max}), \ldots, x^{64}_{\sigma_i}(t_0), \ldots, x^{64}_{\sigma_i}(t_{\max})]^T$$

In order to construct the signal matrix $\mathbf{X}$ which served as input for the PCA, all spatio-temporal signal vectors from the seven BSPMs (σ_i = -75%, -50%, -25%, GG, +25%, +50%, +75%) were combined as follows:

$$\mathbf{X} = \begin{pmatrix} x^1_{-75\%}(t_0) & \cdots & x^1_{GG}(t_0) & \cdots & x^1_{+75\%}(t_0) \\ \vdots & \vdots & \vdots & \vdots & \vdots \\ x^{64}_{-75\%}(t_{\max}) & \cdots & x^{64}_{GG}(t_{\max}) & \cdots & x^{64}_{+75\%}(t_{\max}) \end{pmatrix}$$

Each *column* of $\mathbf{X}$ contained the full spatio-temporal signal vector for one of the seven conductivities under evaluation.

5.6.3.2 PCA Decomposition

The first step during the PCA decomposition is the creation of a mean-free data matrix $\mathbf{X}_{mf}$. To this end, the mean value over all observations is calculated for every row and stored in the m-dimensional vector $\widehat{\mathbf{x}}$. This mean value is subsequently subtracted from all columns of $\mathbf{X}$, which results in the creation of $\mathbf{X}_{mf}$.

The so-called principal components $\mathbf{P}$ are now calculated by solving the eigenvector problem

$$\mathbf{CP} = \mathbf{P}\Lambda \tag{5.7}$$

where $\mathbf{P}$ and Λ are $m \times m$-dimensional matrices and $\mathbf{C}$ is the $m \times m$-dimensional covariance matrix $\mathbf{C} = \mathrm{cov}(\mathbf{X}_{mf})$. The columns $j = 1,\dots,m$ of $\mathbf{P}$ contain the m eigenvectors $\mathbf{p}_j$, and the diagonal elements of Λ contain the corresponding eigenvalues λ_j. All off-diagonal elements of Λ are 0. The magnitude of the eigenvalues λ_j is a measure for the amount of signal variation that is represented by $\mathbf{p}_j$. We used this property to sort the eigenvectors and eigenvalues in order of decreasing eigenvalues.

While it is possible to determine the eigenvectors and eigenvalues as described above, the PCA is often performed more efficiently based on the singular value decomposition [354]. Although this approach delivers only first n-1 eigenvectors, this is usually sufficient for most applications. In this study, we used this method as well by applying an implementation of the modified Golub-Reinsch algorithm [361, 362] from the GNU Scientific Library [363].

5.6.3.3 Signal Reconstruction

If the PCA is interpreted as a coordinate transformation, the calculated eigenvectors span a new orthonormal coordinate system, in which the origin is $\widehat{\mathbf{x}}$ and the base is $\{\mathbf{p}_j \mid j = 1,\dots,m\}$. The spatio-temporal signal vectors $\mathbf{x}_{\sigma_i}$ (*sigma$_i$* represents the seven different conductivity values from -75% – +75%) can now be expressed in the new coordinate system as a superposition of the new base vectors

$$\mathbf{x}_{\sigma_i} = \widehat{\mathbf{x}} + \sum_{j=1}^{m} s_{j,i}\mathbf{p}_j \tag{5.8}$$

The PCA scores $s_{j,i}$ are the entries of the score matrix $\mathbf{S} = \mathbf{P}^T \mathbf{X}_{\mathrm{mf}}$, which is calculated as the projection of the initial mean-free data onto the new coordinate system.

In many applications, the main data variation is already captured in the first few eigenvectors. Thus the PCA can be used to reduce the complexity in data of high dimension as it represents the data more efficiently in its problem specific coordinate system that is formed by the PCA eigenvectors. In this study, it was even sufficient to only consider the first eigenvector, because $\lambda_1 \gg \lambda_2$ as we will show in section 7.6.1. Therefore, signal reconstruction can be done using

$$\mathbf{x}_{\sigma_i} \approx \widehat{\mathbf{x}} + s_{1,i} \cdot \mathbf{p}_1 \tag{5.9}$$

As $\widehat{\mathbf{x}}$ is usually different for the respective tissues, the origin of the coordinate system was shifted to the default signal for the GG conductivities. This was possible because this signal was part of the PCA input matrix for all tissues. This resulted in the shifted scores $q_{1,i} = s_{1,i} - s_{1,GG}$ and the reconstruction formula can be rewritten to:

$$\mathbf{x}_{\sigma_i} \approx \mathbf{x}_{GG} + q_{1,i} \cdot \mathbf{p}_1 \tag{5.10}$$

Although equation (5.9) and (5.10) are mathematically equivalent, equation (5.10) had the advantage that it allowed to combine the results of PCAs from different tissues (see section 5.6.4.2) by using a common coordinate system origin.

5.6.4 Signal Estimation for Arbitrary Conductivities

For a certain tissue and e.g. ventricular input data, the *simulated* BSPM for each of the seven conductivities can be reconstructed from the standard GG signal by adding a certain "portion" of the first principal component (see equation (5.10)). The "size" of this portion for a conductivity σ_i was determined by the shifted score $q_{1,i}$.

In the following sections (5.6.4.1 and 5.6.4.2) we explain how the BSPM can be estimated for an *arbitrary* conductivity that lies between the GG$\pm75\%$ boundaries that were introduced before. This was done for both: Conductivity changes in a single tissue and also for conductivity changes in multiple tissues at the same time.

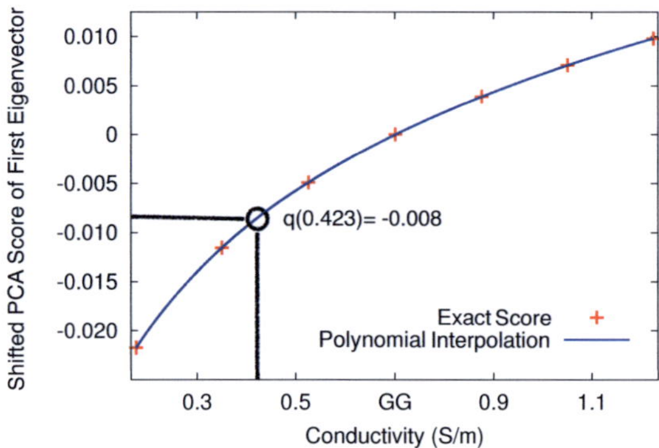

Fig. 5.17. Shifted scores for the PCA decomposition of blood in case of ventricular signals. The PCA coordinate system was shifted so that the GG conductivity lay at the origin (shifted score of the GG conductivity was 0). A polynomial interpolation function could be used to interpolate the associated score for any conductivity between the upper and lower conductivity boundaries. E.g. for $\sigma = 0.423\,\text{S/m}$, the shifted score was $q(\sigma) = -0.008$. The score and the first eigenvector could then be used for the signal reconstruction based on equation (5.11).

5.6.4.1 Signal Estimation for Conductivity Variations in a Single Tissue

Fig. 5.17 shows an example of the shifted scores for the seven blood conductivities in case of ventricular input signals. The scores did monotonically depend on the conductivities. This property was capitalized by fitting a polynomial interpolation function $q(\sigma)$ to the data (range: $\sigma_1 = -75\%$ to $\sigma_7 = +75\%$). This function established a bijective relation between the conductivity σ and the shifted score q. It was now possible to predict $\mathbf{x}$ for any conductivity σ by:

$$\mathbf{x}(\sigma) \approx \mathbf{x}_{GG} + q(\sigma) \cdot \mathbf{p}_1 \qquad (5.11)$$

5.6.4.2 Signal Estimation for Conductivity Variations in Multiple Tissues

When the PCA was performed for the four different tissues (blood, muscle, lungs, fat) the interpolated score curves $q(\sigma)$ and first eigenvectors $\mathbf{p}_1$ were denoted with the respective indices B, M, L, and F. For example in case of blood the score curve would be named $q_B(\sigma_B)$ and the first eigenvector $\mathbf{p}_{1,B}$. Using this naming convention the BSPM for combined conductivity variations could be predicted by:

$$\mathbf{x}(\sigma_B, \sigma_M, \sigma_L, \sigma_F) \approx \mathbf{x}_{GG} + q_B(\sigma_B) \cdot \mathbf{p}_{1,B} + q_M(\sigma_M) \cdot \mathbf{p}_{1,M}$$
$$+ q_L(\sigma_L) \cdot \mathbf{p}_{1,L} + q_F(\sigma_F) \cdot \mathbf{p}_{1,F} \qquad (5.12)$$

This reconstruction equation assumed that it was possible to describe the effects of combined conductivity variations of several tissues as the superposition of signal changes caused by varying the single conductivities separately.

5.6.4.3 Validation of the PCA-based BSPM Prediction

In order to validate our reconstruction approach, which was based only on the first eigenvector of the PCA decomposition, we compared the ratio between first and second eigenvalue. In addition to that, we calculated the pairwise angles between the first eigenvectors of each of the four tissues as these angles were a measure of the independence of changes caused by the different tissues. This was done separately for atria and ventricles.

To assess the errors during signal reconstruction that were due to the omission of the second and successive eigenvectors, we reconstructed BSPM signals for different conductivities (changes in a single tissue) using the *exact* shifted scores and the first eigenvector according to equation (5.10). The root mean square error (RMSE) between the reconstructed and the original forward-calculated signal (which was used as reference) served as quantitative measure for this reconstruction error. The signals which were used as input for the PCA decomposition covered the relevant parts in the cardiac cycle (150 ms of atrial depolarization and 400 ms of ventricular de- and repolarization). Therefore the RMSE was representative for the reconstruction error that can be expected when predicting the BSPM of a typical heartbeat.

In addition to the errors that were associated with our simplified reconstruction equation (5.10), we evaluated the errors that were introduced by the PCA score interpolation based on a leave-one-out validation. To this end, we deliberately omitted an input signal at a specific conductivity from our PCA and used the interpolation technique to derive the associated *interpolated* score. This score was then used to reconstruct the associated BSPM (conductivity changes in a single tissue) and the reconstruction error was assessed by calculating the RMSE. In this case, the RMSE was a measure for the combined errors that were introduced by both the omission of the second and successive eigenvectors and the score interpolation.

Finally, we evaluated the errors associated with the reconstruction of BSPMs that were influenced by simultaneous conductivity variations in all four tissues. In this case, we predicted the BSPMs with all four conductivities either increased or decreased by 25%. This resulted in 2^4=16 combinations, which were reconstructed using equation 5.12. In order to generate reference data that could be used to assess the quality of the reconstruction we performed the associated 16 additional forward calculations using the method described in section 5.6.2. Again the RMSE was calculated between the reconstructed and the reference signals. Here, it was a measure to evaluate the assumption that the combined conductivity variations of several tissues can be predicted by the superposition of the changes introduced by a single tissue separately. A similar validation procedure was repeated for larger conductivity variations of $\pm 50\%$.

5.6.4.4 Confidence Intervals

As explained in section 3.2.3.2, there are multiple reasons for measurement uncertainties of tissue conductivities. However, normally the BSPM changes associated with these uncertainties are neglected as most simulation studies use best-guess conductivities from an arbitrary literature source.

With the presented PCA-based reconstruction approach it is possible to determine the minimum and maximum signal between given conductivity boundaries for each time step. This lower and upper signal boundaries could then be considered as a confidence interval for the reconstructed signal.

In this study, such confidence intervals were calculated based on the best-guess values from GG. Assuming a relative uncertainty $\delta\sigma_{rel}$ the confidence interval was determined for simultaneous uncertainties of $\pm 10\%$, $\pm 30\%$, and $\pm 50\%$ in all four tissues.

Because *bipolar* ECG leads are often used in a clinical setting, we exemplarily evaluated the calculation of confidence intervals in the Einthoven II lead.

5.6.4.5 Conductivity Optimization

During the BSPM signal reconstruction, the PCA uses the relationship between tissue conductivities and PCA scores to predict the effects of changing tissue conductivities. Yet, this relationship is bijective and can also be used in the opposite direction. In this case, it is possible to predict the most likely conductivities for a given BSPM signal. To test this inverse prediction method, we used Nelder-Mead's simplex algorithm [364] as multidimensional optimization scheme to vary the conductivities until a minimal RMSE indicated the best possible match with the available reference signal. A common problem during multi-dimensional parameter optimization is that the search algorithm might get trapped in a local minimum. To prevent this from happening, the optimization procedure was always started ten times from different initial positions. The optimization approach was evaluated for the same 16 reference signals that were used in section 5.6.4.3 (combined tissue conductivities variations in all 4 organs: 2^4 combinations with variations of $\pm 25\%$).

Part III

Results

6

Results: Electrophysiological Modeling

This chapter presents the results of the various *electrophysiological* studies that were conducted during the course of this thesis. The methods that were used during the creation of these results are described in section 4.

6.1 Results: Modeling the Specialized Excitation Conduction System

6.1.1 Effects of the Endocardial Stimulation Profile on the ECG

Fig. 6.1 gives an overview over the effects of changes in the most important stimulation profile parameters on the QRS complex in the Einthoven II lead. The ECG that resulted from the standard parameterization of the endocardial stimulation profile is visualized together with the clinically measured signal in Fig. 6.1A (for details on the standard parameterization or the clinical measurement please refer to section 4.1). The simulated signal showed a similar R peak morphology as the measurement while the S peak was significantly smaller in case of the simulations. QRS complex changes due to a change in conduction velocity within the Purkinje tree are illustrated in Fig. 6.1B. Slower velocities led to a slower depolarization which was associated with a rightwards shifted, broadened R and S peak (of reduced amplitude) and a postponed and prolonged repolarization phase. The effects of an increased conduction velocity were vice versa. A variation in the degree of endocardial coverage with PMJs mainly induced changes of the amplitude and width of the S peak and T-Wave (see Fig. 6.1C). In this case, a lower coverage (30% of the basal surface was uncovered in both left and right ventricle) led to a widening of S peak and T-Wave and increased their amplitudes. In Fig. 6.1D it can be observed that changes in PMJ density had an impact on the amplitude and

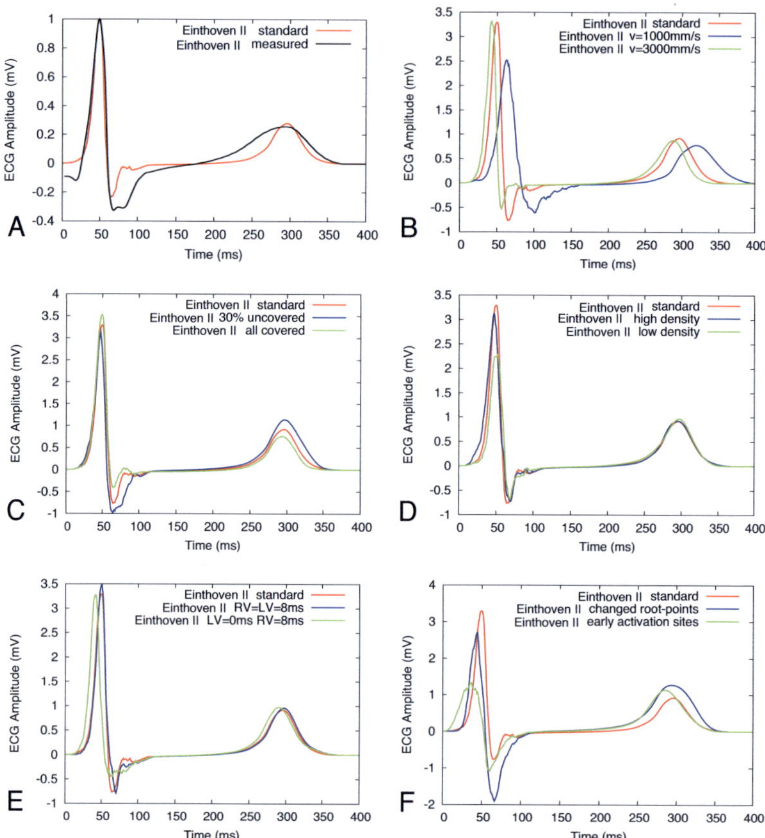

Fig. 6.1. A: Measured and simulated (standard simulation profile) ECG in the Einthoven II lead (normalized on R peak amplitude). B: Changes of the conduction velocity inside the Purkinje tree structure (standard: v = 2000 mm/s). C: Changes of the endocardial surface that is covered with PMJs (standard: left ventricle 15% uncovered, right ventricle 20% uncovered). D: Changes of the PMJ density (standard setting had 744 PMJs; setup with low density had 327 PMJs, setup with high density had 1389 PMJs). E: Changes of the time-offset between left and right ventricle (standard setup: left root-points were stimulated 8 ms after the right root-point). F: Variation of the manually placed root-points and inclusion of additional stimulation points at the sites of early endocardial activation that were reported by Durrer *et al.* [174].

slope of the R peak. A stimulation profile with a lower PMJ density (327 PMJs) generated a low-amplitude R peak with a smaller slope compared to the setup with high PMJ density (1389 PMJs). If we modified the temporal offset between the root-points in the left and right ventricle, this resulted in changes in R and

S peak morphology (see Fig. 6.1E). While changes were small if left and right ventricular root-points were stimulated simultaneously, the R peak was shifted leftwards and the S peak had a reduced amplitude and widened morphology if the stimulation sequence was reversed (in this case the left ventricular root-points were activated first). The effects of a relocation of the root-points can be seen in Fig. 6.1F. Here, the left ventricular root-points have been moved to the anterior and posterior papillary muscles. This resulted in a reduced R peak amplitude whereas the S peak was wider and more pronounced. In addition to that, the T-Wave amplitude was increased and the waveform was widened. Finally, we manually placed additional PMJs at the sites of early left ventricular activation as reported by Durrer *et al.* [174]. This led to both a wider R peak of reduced amplitude (which was additionally shifted to the left) and a wider and deeper S peak. Futhermore, the repolarization occurred faster and the T-Wave amplitude was higher compared to the standard setup.

In Fig. 6.2 the isochrone maps that resulted from the standard stimulation profile are compared with the invasively acquired measurements from Durrer *et al.* [174]. The cut plane was chosen at a similar anatomical position. Although the ventricular anatomy is clearly different, the isochrone maps show many common features (e.g. pattern of early activation in the left ventricle vs. late activation in the posterior part of the right ventricle).

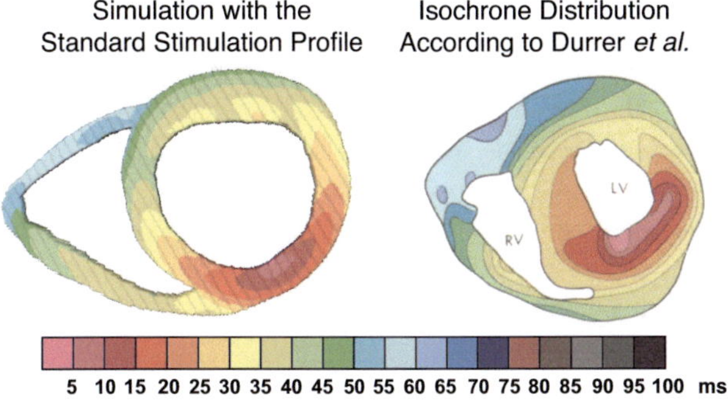

Fig. 6.2. Comparison between the simulated (standard stimulation profile) and a measured isochrone distribution (measurement was conducted by Durrer *et al.*, the figure on the right side is directly adopted from [174]). Both isochrone maps show marked similarities: early activation on the anterior part of the left ventricle and last activation on the posterior side of the right ventricle.

6.1.2 Discussion

Although it is well known that the excitation conduction system is a major determinant for the correct sequence of ventricular activation, there is only a small number of in-silico models available (see section 3.1.4). The main reason for this are contradictory anatomical descriptions, inter-individual variability and the scarcity of human activation-time data. In addition to that, most of the existing models of the excitation conduction system are customized to a special ventricular anatomy and need considerable manual adjustments in order to be transferable to a new dataset. In this study, we present a method that allowed the fast creation of an endocardial stimulation profile that was designed to mimic the specialized excitation conduction system. Through variable model parameters it was possible to rapidly customize the stimulation profile or to adapt it to different ventricular anatomies.

In addition to that, we varied the different parameters and evaluated their impact on the Einthoven II lead. In this context it is interesting to note that parameters which had only an impact on the height of the R and S peak like the density of the PMJs (Fig. 6.1D) or the time offset between the root-points (Fig. 6.1E) induced no changes of the T-wave morphology. In contrast to that changes of the width of the R and S peak due to modifications of the conduction velocity (Fig. 6.1B), endocardial coverage (Fig.6.1C) or the manually placed root-points (Fig. 6.1F) impacted directly on the width and height of the T-wave.

If the left ventricular root-points were stimulated 8 ms after the one in the right ventricle, the QRS complex showed a better match with the clinical recording. This was unexpected as Durrer *et al.* claim that the first ventricular activation occurs in the left ventricle [174]. Furthermore, the inclusion of Durrer's sites of early activation did not enhance the match with the clinically acquired ECG (see Fig. 6.1F).

In conclusion, we presented a versatile method that allowed the fast and realistic creation of endocardial stimulation profiles. To further facilitate the customization and adaption to new ventricular anatomies, we envision an optimization framework that will automatically determine an optimal parameterization of the stimulation profile and thus further reduce time consuming manual interaction. Such an optimization framework will be implemented and evaluated in a future study.

6.2 Results: Electrophysiological Heterogeneities in the Ventricles

The measured and a simulated ECG (**TM-40*A$^{1.5}$B^1**) are visualized for comparison in Fig. 6.3A and Fig. 6.3B. The signal amplitudes of the clinically acquired ECG depend on instrumental boundary conditions like the utilized electrodes, electrode gel and internal signal amplification. As these parameters will change with a different ECG measurement system we chose to discard this information by normalizing all measured and simulated signals (Fig. 6.3A and Fig. 6.3B, Fig. 6.4A-D) to their respective R-peak amplitude in the Einthoven II lead. The simulated activation sequence was identical in all setups. A good agreement was reached between the simulated and measured QRS complex (compare Fig. 6.3A and Fig. 6.3B). In case of the R-peak, the amplitude was smallest and the peak showed a leftwards shift in Einthoven I whereas a rightwards shift could be observed in Einthoven III. R-peak amplitude was largest in Einthoven II, which also had a prominent S-peak.

In Fig. 6.4 the Einthoven II lead of all heterogeneous g_{Ks} configurations is displayed (they are described in detail in section 4.2.3). The signals were sorted according to the implemented heterogeneities: transmural heterogeneities (Fig. 6.4A), apico-basal and interventricular heterogeneities (Fig. 6.4B), combined apico-basal and transmural heterogeneities (Fig. 6.4C) and other electrophysiological configurations (Fig6.4D). The plots of these signals give a visual impression of the signal morphology compared to the reference measurement.

However, in order to evaluate the whole spatio-temporal pattern of the T-Wave, a simple visual evaluation is not sufficient as it is always focused on specific leads thereby neglecting possible changes elsewhere. A more quantitative analysis that

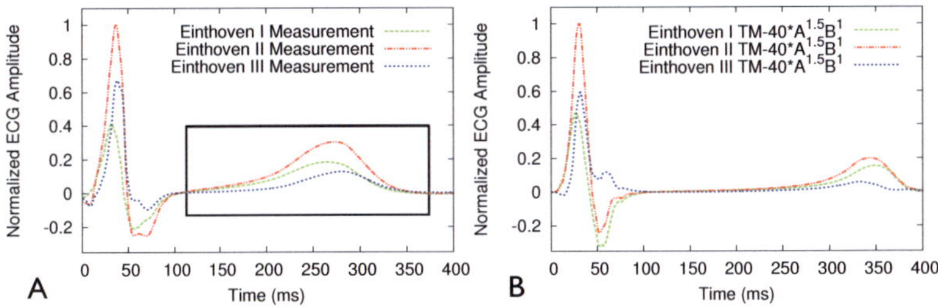

Fig. 6.3. Measured (A) and simulated (B) ECG in the Einthoven leads. A good agreement was reached between the simulated and measured QRS complex. The box that is visualized on top of the measured signal indicates the window that was used to crop a part of signal around the T-Wave for further evaluation.

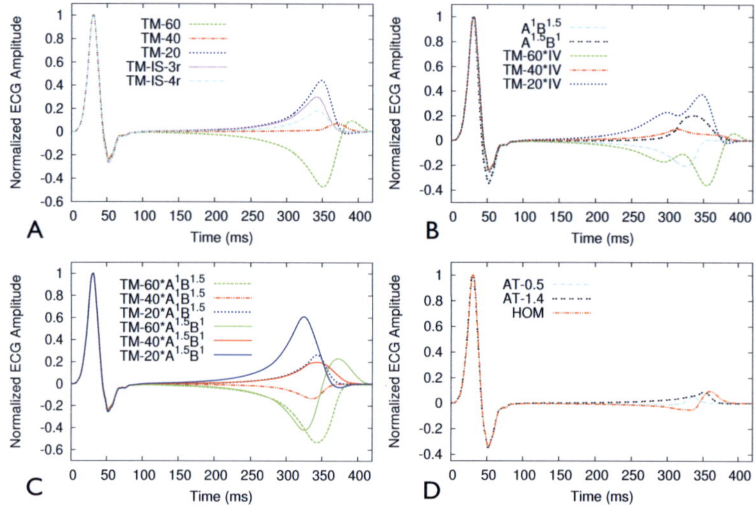

Fig. 6.4. Einthoven II lead for the different heterogeneous setups (normalized to the respective R-peak maxima). A: setups with transmural heterogeneities, B: setups with apico-basal and interventricular heterogeneities, C: mixed apico-basal and transmural heterogeneities, D: other electrophysiological configurations as explained in section 4.2.3.5.

considers global BSPM pattern changes is based on the the averaged correlation coefficient $\bar{r}$ and the averaged root mean square error (RMSE) between the measured T-Wave and the T-Waves of the different simulations (see Fig. 6.5). Both correlation coefficient and RMSE were averaged over all electrodes that were present in the post processed multi-channel ECG recording. As the QT time was different in each of the simulations and therefore also between the measurement and the simulations, it was not possible to calculate the averaged correlation coefficient and RMSE over the whole signal length due to the associated differences in T-Wave position. To circumvent this problem, we calculated both evaluation criteria in a window that contained only a cropped part of the signal (i.e. the T-Wave and parts of the ST segment).

In order to be able to compare the cropped signals from the measurement and the simulations, we had to ensure that the cropped signals were extracted from similar time instants during the cardiac cycle. This was done by finding the window position that delivered the maximum averaged correlation between the measured

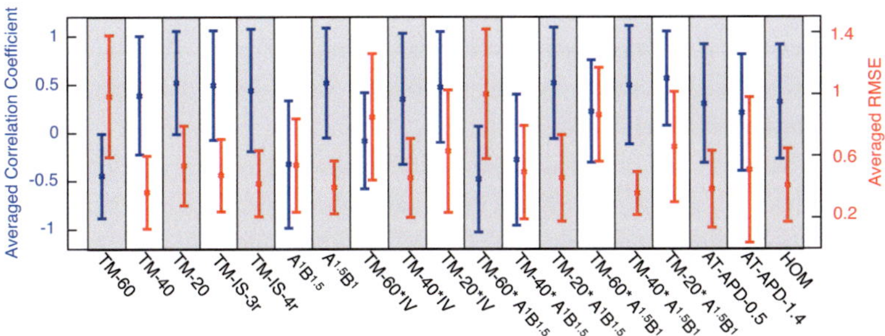

Fig. 6.5. Averaged correlation coefficient and averaged and normalized RMSE between the measured and simulated signals. Both criteria were calculated within a signal window (with 261 ms length) that contained the T-Wave and parts of the ST segment. This had the advantage that the evaluation was possible regardless the differences in QT times between the various setups.

and the simulated signal. More details on this approach will now be given in the following:

At first, we had to define a window for the measured ECG signal. To this end, we determined the T-Wave maximum in an arbitrary lead with a positive T-Wave and chose the window to start 160 ms before this maximum and to end 100 ms after it. The resulting window with a length of 261 ms is visualized in Fig. 6.3A.

Now we had to find the corresponding window for each simulated signal. In this case, we again determined the T-Wave maximum in an arbitrary positive lead and used it as an initial guess for the window position. For this position, we then calculated the correlation coefficient between every channel of the measured and simulated signals. Based on these channel specific correlation coefficients it was possible to calculate the averaged correlation coefficient $|\bar{r}|$ by:

$$|\bar{r}| = 1/n \sum_{i=1}^{n} |r_n| \qquad (6.1)$$

where n was the channel number.

Within these windows, the average correlation coefficient $\bar{r}$ was calculated between each heterogeneous setup and the measured signal by:

$$\bar{r} = 1/n \sum_{i=1}^{n} r_n \qquad (6.2)$$

This averaged correlation coefficient is visualized together with the standard deviation over the electrodes of each heterogeneous setup in Fig. 6.5. High morphological similarities are indicated by large values of $\bar{r}$. Values below 0 denote T-Waves with opposite polarities.

Although the correlation coefficient is a suitable measure to compare different signal morphologies, it neglects signal amplitudes and is therefore not able to evaluate if the amplitude relations between QRS complex and T-Waves are realistic in the simulated signals. For this reason, we chose to analyze these amplitude relations by calculating the averaged RMSE. Prior to the calculation of the RMSE all signals were normalized to their respective R-peak amplitude in the Einthoven II lead. The same signal windows that were previously used to determine the averaged correlation coefficient $\bar{r}$ were now used to calculate the RMSEs between measured and simulated signals. The RMSEs over all channels were subsequently averaged and normalized using the largest averaged RMSE of all setups (in this case: $\mathbf{TM\text{-}60\text{*}A^1B^{1.5}}$). The resulting values are plotted together with the standard deviation in Fig. 6.5. All heterogeneous configurations that had both a large value of $\bar{r}$ and a small averaged RMSE showed a good agreement with the measured reference T-Wave.

In experimental studies, the repolarization time (RT) is an often determined parameter that is used to characterize the repolarization sequence. It is defined as the sum of activation time and action potential duration ($RT = AT + APD_{90}$). In Fig. 6.6 and Fig. 6.7 the repolarization time maps of all 19 heterogeneous configurations are visualized.

In general, it was not possible to rule out the existence of either apico-basal or transmural heterogeneities as setups with both, exclusively apico-basal and exclusively transmural heterogeneities resulted in concordant T-Waves. However there were some transmural and also apico-basal configurations that seem to be unlikely to generate positive T-Waves. With respect to transmural heterogeneities it could be observed that a more endocardial position of the M-cells led to positive T-Waves of large amplitudes. In this case, the repolarization started from the epicardium (see Fig. 6.6). If the M-cells were moved further to the epicardium, the repolarization started from the endocardium and the T-Wave was inverted.

If we look at the effects of apico-basal heterogeneities, it could be observed that a higher apical density of I_{Ks} resulted in positive T-Waves with repolarization start-

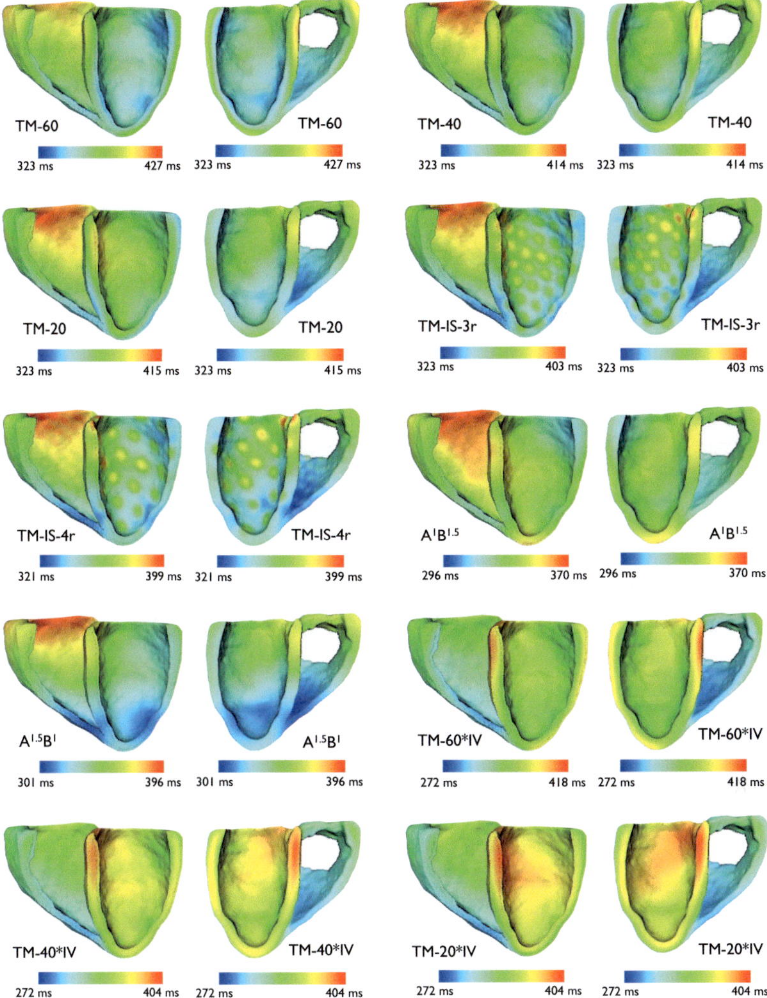

Fig. 6.6. Repolarization time maps for the setups with transmural, apico-basal and interventricular heterogeneities. Please note the individual color-scaling of each repolarization time map.

ing at the apex. Negative T-waves were seen for higher basal I_{Ks} densities. In this case, the repolarization started at the base of the LV and at the apex of the RV.

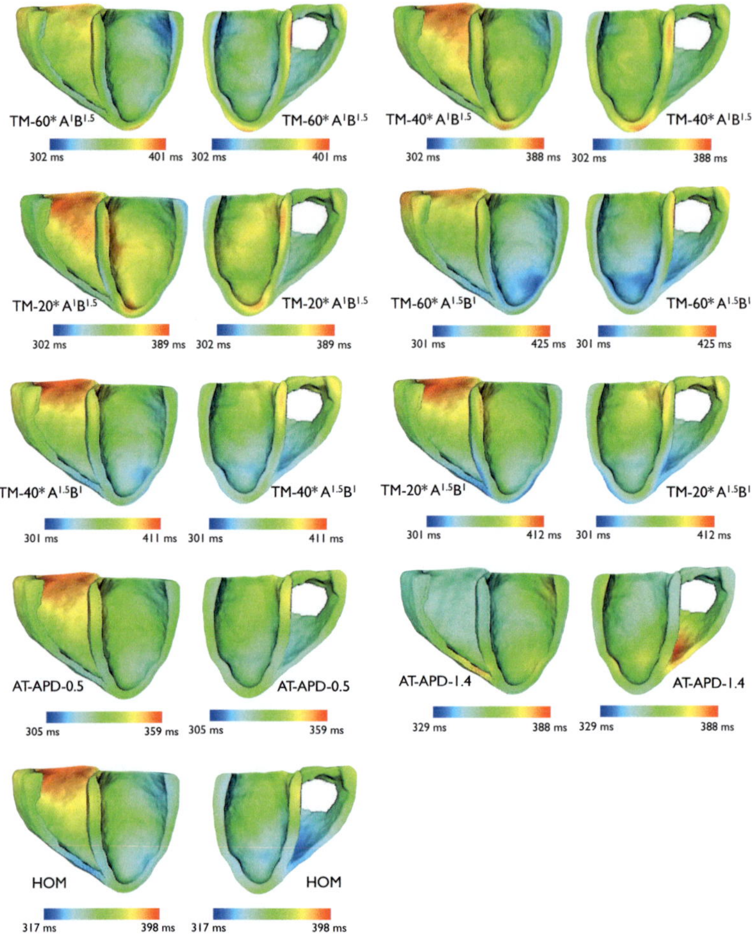

Fig. 6.7. Repolarization time maps for setups with a combination of apico-basal and transmural heterogeneities and setups that had a linear relationship between AT and APD. The repolarization time map in the last row visualized the repolarization sequence in a homogeneous setup. Please note the individual color-scaling of each repolarization time map.

If interventricular heterogeneities were added to transmural heterogeneities, it always resulted in notched T-Waves. Due to the higher I_{Ks} density, the repolarization began in these setups at the RV apex. The combination of transmural and apico-basal setups produced positive, negative and biphasic T-Waves. Again a more en-

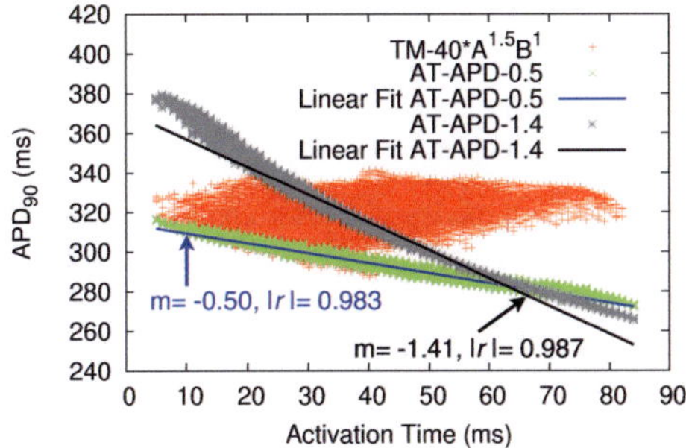

Fig. 6.8. Relation between AT and APD for three different heterogeneous setups. In two cases, this relation was linear. In that case, the linear fit is given together with the corresponding correlation coefficient. The third case had a more complex relation between AT and APD. Despite this non-linear relationship, this setup produced a concordant T-Wave.

docardial location of the M-cells and larger apical I_{Ks} densities tended to produce larger concordant T-Waves.

Finally, both setups that had a linear relationship between AT and APD had positive T-Waves although they were of low amplitude. A comparable signal shape was also seen for the setup with homogeneous channel densities. The relationship between AT and APD was further characterized for three selected signals (**AT-APD-0.5**, **AT-APD-1.4** and **TM-40*A$^{1.5}$B^1**) in Fig. 6.8. For both signals with linear relationships between AT and APD, the linear fits are given together with the corresponding correlation coefficients r. Due to the fact that all three setups in Fig. 6.8 have concordant T-Waves but not all setups had linear relationships between AT and APD, it became evident that such a relationship was not a prerequisite for concordant T-Waves.

6.2.1 Discussion

Although it is commonly accepted that the APD is distributed heterogeneously throughout the ventricles and that the associated dispersion of repolarization is mainly responsible for the shape of the T-Wave, its concordance and exact morphology are still not completely understood.

In this context, quantitative electrophysiological modeling of the heart is deemed to be important as such in-silico models allow to predict the effects of changes in individual ion channels on the body surface ECG. To date there are some software packages available (e.g. ECGSIM [365]) that allow the fast and realistic calculation of ECGs. However these models rely on simple electrophysiological descriptions and do not allow to integrate experimentally measured ion channel heterogeneities.

In this study, we strove to fill this gap by using an anisotropic heart and torso model in combination with a detailed model of cardiac electrophysiology to determine the effects of different heterogeneous I_{Ks} distributions on the ECG. In contrast to other studies that investigated the effects of repolarization heterogeneities using a cellular automaton approach [159, 160, 161], we simulated action potential propagation with the monodomain reaction diffusion equations which has several advantages over rule-based fixed AP models as outlined in [162].

In order to be able to create representative I_{Ks} distributions for a subsequent evaluation, we compiled a literature overview concerning heterogeneous ion channel densities and APD dispersion (see Table 4.1, Table 4.2 and Table 4.3). Although this list is not complete, it represents a collection of studies that are often used in the modeling community as basis for models of heterogeneous ion channel distributions. In addition to that, the tables also illustrate the conflicting reports between measurements on different, or even within the same species (e.g. direction of apico-basal dispersion).

It should further be noted that the existence and effects of the M-cells is controversially discussed in the community. Although we listed a number of studies that found M-cells in a variety of species, there is also a large body of reports that disputes their existence or at least their functional role regarding the genesis of the T-Wave. The list of species in which the existence of M-cells is questioned comprises canine [366, 367, 368], pig [369, 370], rat [371] and human [111, 372, 373]. The contradictory findings are mainly attributed to differences in measurement techniques [67, 374] and the use of certain anesthetic agents [374]. Other factors that might influence the development of characteristic M-cell features in cardiac tissue is the age of the animals under study [375] and the type of sample on which the measurements are conducted (wedge vs. intact heart) [376].

In this study we did not try to prove or refute the existence of the M-cells. We rather aimed at a neutral evaluation of the effects of previously measured heterogeneities on the morphology of the T-Wave. To this end, we evaluated not only transmural but also apico-basal, interventricular and mixed setups. It should be noted, that none of the tested setups was able to reproduce the measured T-Wave in the multichannel ECG completely. However, there were some configurations that appeared to be more likely and others that could be ruled out as they produced discordant T-Waves.

Examples for setups that performed above average were the transmural configurations in which the M-cells were placed in Mid position (**TM-40**) or had island-shaped topography (**TM-IS-3r, TM-IS-4r**). In contrast to that, a more epicardial M-cell position as proposed by Drouin *et al.* [261] is unlikely as it was not able to generate a concordant T-Wave. Furthermore, our model did also not support the existence of a higher basal density of I_{Ks} (see Fig. 6.4 and Fig. 6.5). The best match between measured and simulated T-Waves was achieved for the apico-basal configuration with higher apical I_{Ks} density ($\mathbf{A^{1.5}B^1}$) and a mixed transmural and apico-basal configuration (**TM-40*$\mathbf{A^{1.5}B^1}$**).

If we look at the setups that had an inverse linear relationship between AT and APD, it was surprising to see that both tested slopes (-0.5 and -1.4) produced concordant T-Waves. This was not expected as it challenged the rule of thumb according to which the repolarization should travel in the opposite direction than the depolarization to generate concordant T-Waves. However, these findings were supported by the study from Yuan *et al.* [292] where slopes < -1 were found in the presence of positive T-Waves. Finally it should be noted that the T-Waves resulting from both setups had unrealistic signal amplitudes (see Fig. 6.5) which renders the associated g_{Ks} distributions to be rather unlikely.

In general, it was difficult to predict T-Wave morphologies based on the repolarization time (RT) maps that are displayed in Fig. 6.6 and Fig. 6.7. This is particularly evident if we compare the RT map of **TM-60** with the map of the **HOM** setup. Despite the fact, that both AT maps are similar and that they show a repolarization sequence that is comparable to the activation sequence (compare with Fig. 4.3), both setups generate T-Waves of different polarities. This emphasizes the importance of quantitative cardiac modeling as it might not be sufficient to perform a (low) number of isolated RT measurements to reliably predict the T-Wave morphology.

Our study is limited by the fact, that the sequence of repolarization is not only determined by the APD but also by the sequence of activation. This means that a different sequence of activation would lead to differences in the sequence of repolarization and thus to different T-Wave morphologies. We tried to minimize this potential source of error by modeling an activation sequence which resembled the real activation sequence as much as possible (compare measured and simulated

QRS complex in Fig. 6.3). In addition to this dependence of the repolarization sequence on ventricular activation, the T-Wave morphology might also depend on the shape of the anatomical model that is used for the investigation. In case of an anatomical model that has thicker ventricular walls, the depolarization front needs more time to travel through the wall and thus the model will react differently to certain distribution of transmural or even apico-basal heterogeneities [377]. Finally we used a static diastolic model of the ventricles that neglected ventricular contraction and relaxation. The consideration of these effects has been shown to lead to a shortened QT time [260] and a reduced T-Wave amplitude [13, 260]. These effects should be kept in mind when interpreting the results of this study.

When it comes to the formation of the T-Wave, it is known that not only the length and time of initiation of an action potential is important, but also the height and slope of the action potential plateau (see response in [372]). This means that a prominent T-Wave can exist even if the differences in repolarization times are small. In addition to that, differences in action potential slope are probably also responsible for the constant rise in signal amplitude in the ST segment of the measured ECG (see Fig. 6.3A). As this was not visible to the same extent in case of the simulated signals, we conclude that our model had a lower than normal heterogeneity concerning different action potential morphologies. This issue should be addressed in future studies by the inclusion of additional ion channel heterogeneities that impact on the shape rather than the duration of the ventricular action potential.

In conclusion, we presented an evaluation that used a realistic in-silico model of the ventricles to investigate the effects of different ion channel heterogeneities on the morphology of the T-Wave. The results were compared to a clinical ECG recording from the same volunteer that delivered the anatomical basis for the simulation study. To the best of our knowledge, this is the first biventricular and electrophysiologically detailed model that allows to assess the influence of ion channel heterogeneities by comparing the results with real clinical ECG data. This model can be used in the future whenever new measurement data becomes available to evaluate its plausibility with respect to the formation of concordant T-Waves.

6.3 Results: Beta-Adrenergic Regulation of Cellular Electrophysiology

6.3.1 Calcium Sparks and Adrenergic Influence on I_{NaK}

In our initial implementation of the beta-adrenergic signaling pathway, we considered its effects only on three target proteins: I_{CaL}, SERCA (I_{up}) and I_{Ks} [5, 7]. After conducting some test simulations for longer simulation durations (up to 60 s), it soon became obvious that the model (especially with M cell parameterization) was prone to spontaneous calcium release currents (from now on termed calcium sparks) from the SR which could trigger early after depolarizations (EADs) as can be seen in Fig. 6.9.

The reasons for these calcium sparks was the cytosolic accumulation of calcium that was evoked by the increase of I_{CaL} under beta-adrenergic activity. In addition to that, also more calcium was present in the SR due to the ISO-induced increase in I_{up}.

As a first reaction, to counteract these calcium sparks, we reduced the ISO-effects on I_{CaL} by adapting the phosphorylation coefficients as described in [5, 7]. However, after conducting a literature survey it became evident, that the existence of calcium sparks was normal if the adrenergic regulation on I_{NaK} was not considered [299].

Despa *et al.* conducted experiments investigating the effects of ISO on I_{NaK} and its indirect influence on the cytosolic calcium concentration via NCX [299]. Concerning I_{NaK}, a small single-membrane spanning protein called phospholemman (PLM) seems to be responsible for the regulation of the pump activity. In resting

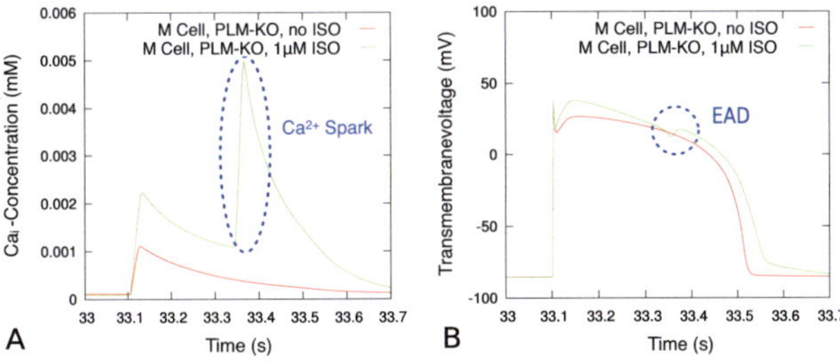

Fig. 6.9. Example for a spontaneous calcium release current from the SR (shown here is an M cell configuration that was paced at 1 Hz). The calcium spark triggered an early after depolarization (EAD) that might become arrhythmogenic. Figure was modified based on [6].

Table 6.1. Number of calcium sparks that occurred for the different cell types (endo, M and epicardial cell) under different stimulation frequencies (0.5 Hz, 1 Hz, 2 Hz) in the presence of ISO during a simulation duration of 60 s. Two different in-silico model versions are compared: a model without ISO-effects on I_{NaK} (PLM-KO) and a model that considers these effects (WT).

	Endo			M			Epi		
	0.5 Hz	1 Hz	2 Hz	0.5 Hz	1 Hz	2 Hz	0.5 Hz	1 Hz	2 Hz
PLM-KO	0x	1x	6x	0x	11x	0x	0x	6x	12x
WT	0x	0x	0x	0x	7x	0x	0x	0x	3x

conditions, PLM inhibits I_{NaK} by reducing its affinity for cytosolic sodium. In the presence of ISO, PLM is phosphorylated thereby inducing a leftwards shift in the sodium activation curve (i.e. the inhibiting effect is reduced). In their investigations, Despa *et al.* used healthy mice (WT) and PLM knockout mice (PLM-KO) in which the phosphorylation of PLM was no possible. WT mice showed a smaller ISO-induced increase in both cytosolic and SR calcium content due to the indirect effects of I_{NaK} on NCX that were explained previously (see section 4.3.1) [299].

To consider the described effects, we integrated the adrenergic effects on I_{NaK} into our model as well (see section 4.3.2). As initial parameterization, fac_{kNaK} was set to 0.011 and $fac_{K_{mNa}}$ was set to 0.34. To compare the susceptibility of this updated model towards the development of calcium sparks to the original version from [5, 7], the following in-silico experiments were conducted: Endo, M and epicardial configurations of the model were stimulated at frequencies of 0.5 Hz, 1 Hz and 2 Hz for a total duration of 60 s. Then, the number of calcium sparks occurring within this period was counted. The results were sorted depending on whether adrenergic effects were considered to regulate I_{NaK} (WT) or not (PLM-KO). Table 6.1 summarizes the results:

None of the three different cell types showed calcium sparks for stimulation rates of 0.5 Hz both with (WT) or without ISO effects on I_{NaK} (PLM-KO). For a stimulation rate of 1 Hz, all cell types showed sparks in the PLM-KO configuration (highest incidence in M-cells) which disappeared in endo- and epicardial cells once the influence of ISO on I_{NaK} was considered. In case of the M cells, the number of sparks was reduced in the WT configuration. At higher stimulation rates (2 Hz), only endo- and epicardial cells showed calcium sparks (PLM-KO). In case of endo cells, the sparks disappeared and in case of epicardial cells, they occurred less frequently once ISO effects on I_{NaK} were considered.

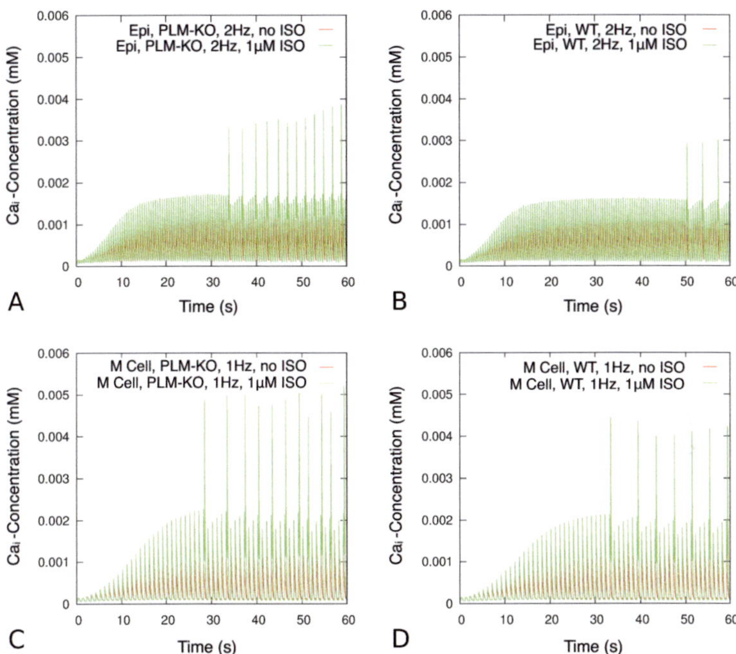

Fig. 6.10. Example for the reduction in calcium spark occurrence if ISO effects on I_{NaK} are considered. A: Epicardial cells paced at 2 Hz without ISO effects on I_{NaK} (PLM-KO) showed 12 sparks in the presence of ISO. B: In contrast to that, epicardial cells paced at 2 Hz with ISO effects on I_{NaK} (WT) showed only 3 sparks in the presence of ISO. The same trend was seen for M cells paced at 1 Hz. Here, the number of sparks was reduced from 11 (C) to 7 (D). Figure was modified based on [6].

Fig. 6.10 illustrates the results from Table 6.1 for two examples: In case of epicardial cells that were stimulated with 2 Hz, the number of sparks was reduced from 12 (PLM-KO) to 3 (WT) if ISO effects on I_{NaK} were considered. Similarly, sparks occurred less frequently in case of M cells which were stimulated at 1 Hz (11 sparks for PLM-KO vs. 7 sparks for WT).

In the following, the influence of ISO on I_{NaK} was increased in order to evaluate the potential effects on the occurrence of the calcium sparks. To this end, both fac_{kNaK} and $fac_{K_{mNa}}$ were increased from their WT values. In case of "WT x 1.5" both variables were multiplied by 1.5 (corresponding multiplications were denoted by "WT x 2" and "WT x 2.5"). The effects were analyzed for M cells stimulated with 1 Hz (see Fig. 6.11). In general, the following rule applied: the larger the

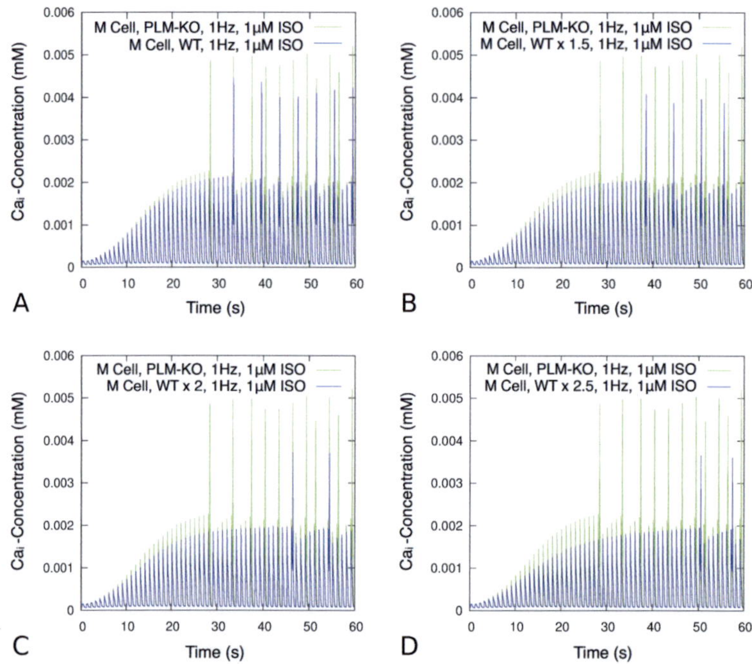

Fig. 6.11. Effect of increased strength in beta-adrenergic influence on I_{NaK}. The factors that regulate conductivity and affinity to sodium were multiplied by 1.5 (B: WT x 1.5), 2 (C: WT x 2) and 2.5 (D: WT x 2.5). As a consequence to the increased ISO effects, the number of sparks declined. Figure was modified based on [6].

ISO-induced increase of I_{NaK}, the lower the cytosolic calcium concentration and the lower the number of sparks that were visible.

In order to further tune the weighting of the parameters that are responsible for the beta-adrenergic effects on I_{NaK}, we reproduced an experiment that was originally conducted by Despa *et al.* [299]. In this experiment, myocytes from WT and PLM-KO mice were paced at 2 Hz while the cytosolic sodium concentration was monitored (see Fig. 6.12A). After 7 or 8 minutes of pacing, ISO was administered. In WT mice, in which an intact beta-adrenergic regulation of I_{NaK} was present, the application of ISO reduced the cytosolic sodium concentration by increasing the pump activity of I_{NaK}. In PLM-KO mice, this effect was not present and the sodium concentration was unaffected by ISO. In the in-silico experiments, the ef-

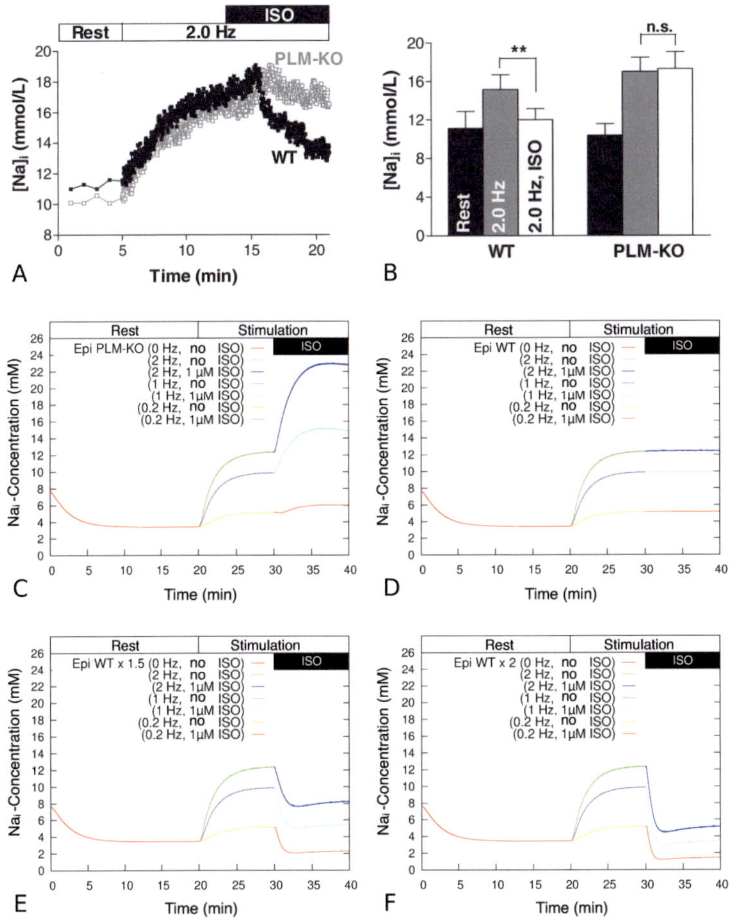

Fig. 6.12. Experimental results adopted form the study from Despa *et al.* [299] (A,B). A: Single experiment in which the course of the cytosolic sodium concentration can be observed over time. B: Pooled data from several experiments. Shown are the concentrations at the end of the resting period (Rest), just before the administration of ISO (2.0 Hz) and at the end of the recording (2.0 Hz, ISO). C-F: Cytosolic sodium concentration in epicardial cells with different weightings of the ISO influence on I_{NaK}. The larger the influence, the faster the decay of Na_i. Figure was modified based on [6].

fect of different strengths of the ISO influence on I_{NaK} was tested. The results can be summarized as follows:

- Without adrenergic effects on I_{NaK}, the cytosolic sodium concentration rose in the presence of ISO (see Fig. 6.12C). This did not agree with the experimental results from Despa *et al.* [299]. However, it should be noted that they were surprised by the fact that ISO had no effect on Na_i in PLM-KO mice in their experiments. Actually they expected to see a rise in cytosolic sodium concentration in this setting (similarly as it was observed in the simulations). The reason for this was the larger calcium influx that was party balanced by the enhanced calcium extrusion via NCX. Due to the effects on I_{NaK}, this should in turn increase cytosolic sodium as explained in [299].

- The larger the phosphorylation effects on I_{NaK}, the stronger the decay in cytosolic sodium: For the WT setup, no change of Na_i was observable. However, for an increased phosphorylation (factor 1.5 or 2) similar reductions were seen as in the experiments.

Based on these results, one could argue that the phosphorylation effects on I_{NaK} are not adequately represented in the WT setup. However, additional experiments should be conducted (e.g. the time-course of cytosolic calcium could be reproduced based on [299]) before final adaptations are made.

6.3.2 Effects of Beta-Adrenergic Regulation on Endo, M and Epicardial Cells

In the following, the effects of an activated beta-adrenergic signaling pathway are shown on endo, M and epicardial cells (see Fig. 6.13, Fig. 6.14, Fig. 6.15). The main effects on the four channels in which ISO influence was considered were already described in section 4.3.2. Here, graphical visualizations of these changes are shown for the sake of completeness.

6.3.3 Changes of AP Morphology Due to Beta-Adrenergic Regulation

After including the beta-adrenergic signaling pathway into the ten Tusscher model, the repolarizing flank of the AP showed a small slope just before the repolarization was finally completed (see Fig. 6.16A). Such a change in AP morphology was neither observed in the original Saucerman model [100] nor in experimental measurements on a ventricular wedge preparation (see Fig. 6.16B). When conducting in-silico experiments, this change in repolarization slope could severely affect AP evaluation criteria like the APD_{90}. Such problems were encountered e.g. in a study

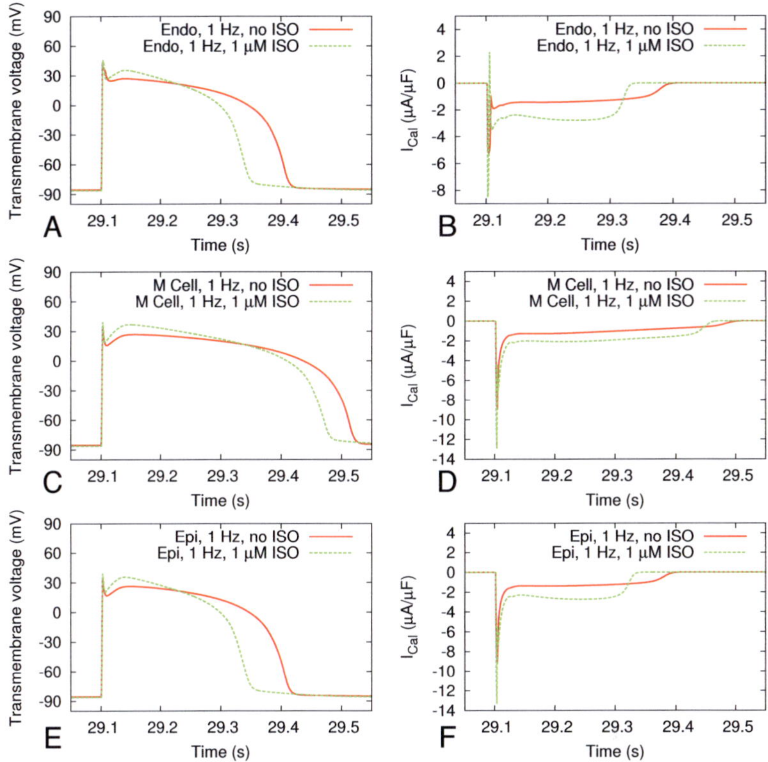

Fig. 6.13. A,C,E: Visualization of the AP shape for endo, M and epicardial cells in the presence and absence of ISO. If ISO was administered, the APD was shortened in all three cell types. However, the shortening was not uniform: endo- and epicardial cells were shortend to a greater extent than M cells. This could give rise to an increase in dispersion of repolarization in the presence of ISO due to the larger differences in APD. B,D,F: Visualization of I_{CaL} in endo, M and epicardial cells in the presence and absence of ISO. If ISO was administered, the maximum amplitude of the calcium current as well as the plateau component increased. In endocardial cells, the calcium current is reversed (for a few ms) shortly after reaching its maximum. Although this effect had no physiological foundations, it did not seem to have a significant influence on the calcium content of the SS or cytosol.

were the APD_{90} was used to calculate changes in TDR that were associated with the LQTS in the presence and absence of ISO (see section 6.4.3).

To determine the phosphorylated ion channel that was responsible for this change in AP morphology, the following approach was chosen: The ISO-effects on each channel were removed separately and the AP shape was then analyzed. Only trans-

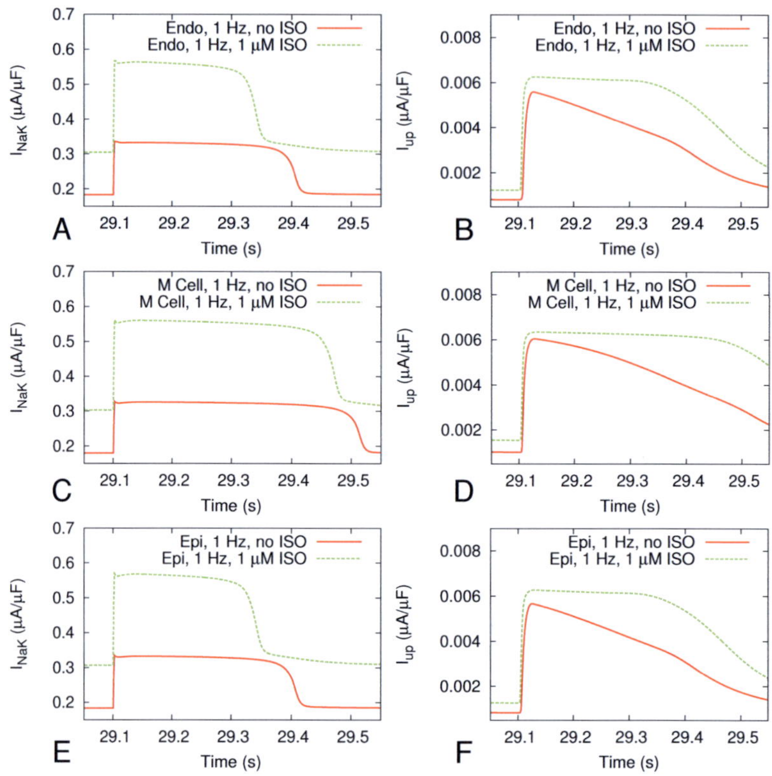

Fig. 6.14. A,C,E: Visualization of I_{NaK} in endo, M and epicardial cells in the presence and absence of ISO. The activity of the pump both during the course of the AP as well as in resting conditions was enhanced under the influence of ISO. B,D,F: Visualization of I_{up} in endo, M and epicardial cells in the presence and absence of ISO. If ISO was administered, more calcium was pumped back into the SR.

membrane currents were considered in the evaluation. I_{Ks}-KO denoted a model without ISO effects on I_{Ks} whereas PLM-KO and I_{CaL}-KO represented models without phosphorylation of I_{NaK} and I_{CaL}, respectively.

Fig. 6.17 summarizes the results of this experiment. It is obvious that the ISO effects on the slope of the repolarizing flank were associated with I_{CaL} (see Fig, 6.17C). For future revisions of the beta-adrenergic signaling pathway, the implementation of I_{CaL} could be improved. As a matter of fact, there is a study from

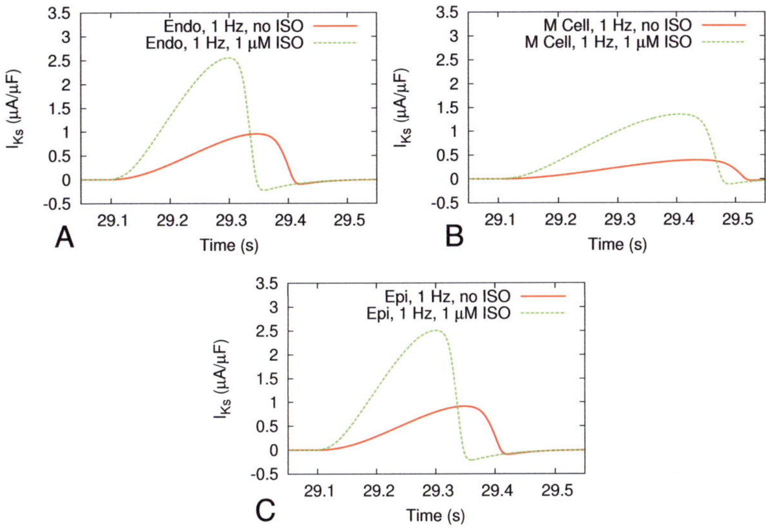

Fig. 6.15. A,B,C: Visualization of I_{Ks} in endo, M and epicardial cells in the presence and absence of ISO. The ISO-induced increase in I_{Ks} amplitude was responsible for the shortening of the APD under beta-adrenergic influence.

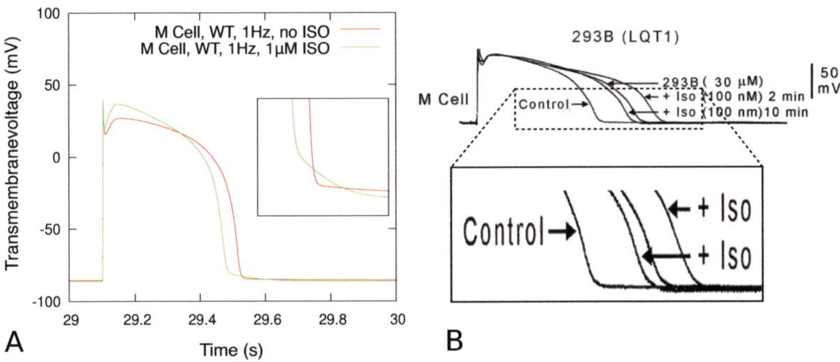

Fig. 6.16. A: Changes in AP morphology are shown for an M cell configuration in the presence of ISO (see inlet for details). B: For comparison AP traces from a ventricular wedge experiment are shown. Here, no changes concerning the slope of the repolarizing flank were visible [221]. Figure was modified based on [6].

Chen *et al.*, which provides data on human I_{CaL}, even considering ISO effects on the channel [378].

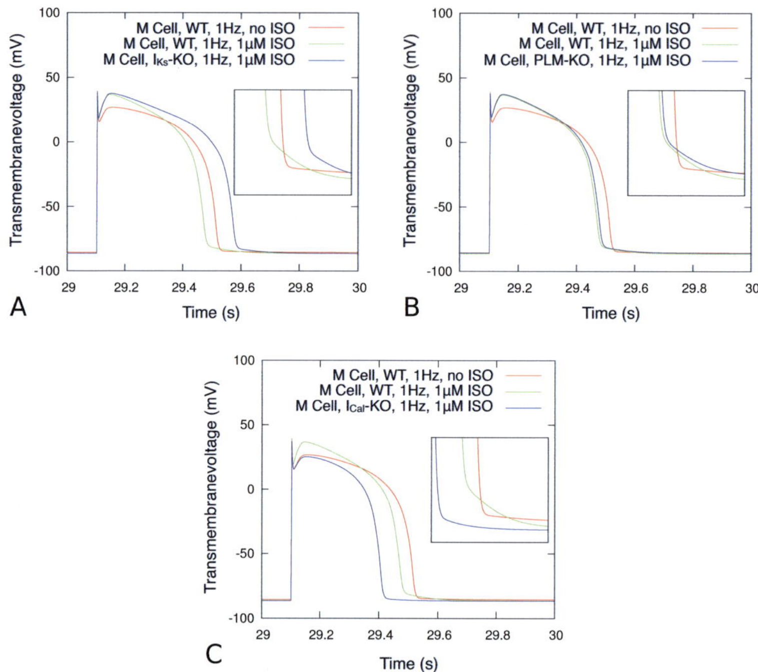

Fig. 6.17. A: Effects of the removal of ISO influence on I_{Ks} (I_{Ks}-KO). In this case the APD was prolonged compared to the APD without ISO. This was due to the stronger influx of calcium through the phosphorylated I_{CaL} which prolonged the plateau phase. Yet no changes on the slope of the repolarizing flank were visible. B: Effects of the removal of ISO influence on I_{NaK} (PLM-KO). No significant changes of the AP morphology were observable. C: Effects of the removal of ISO influence on I_{CaL} (I_{CaL}-KO). Here, the APD was shortened significantly. This was due to the fact that ISO influence was still present on I_{Ks} (increase of the repolarizing slow potassium current). Yet ISO influence was removed from I_{CaL} which means that no additional calcium was flowing into the cell. For this configuration, the repolarizing flank of the AP looked similar to flank of the original ten Tusscher model. Therefore it can be concluded that the phosphorylation of I_{CaL} is responsible for the observed morphological changes. Figure was modified based on [6].

6.4 Results: The Congenital Long-QT Syndrome

6.4.1 Results: Multiscale Modeling of Long-QT 2 in the Visible Man Torso: A Feasibility Study

6.4.1.1 LQT 2 Effects on APD and ECG

The influence of LQT 2 (in different severities) on electrically coupled endocardial, M and epicardial cells is illustrated in Fig. 6.18. As LQT 2 was modeled by a reduction of I_{Kr} channel conductivity, we observed a prolongation of the APD_{90} in all three cell types. Under physiological conditions, the heterogeneous distribution of I_{Ks} leads to a longer APD_{90} in M cells (as the I_{Ks} density is considerably lower here compared to endo- and epicardial cells): $APD_{90,Endo} = 318$ ms, $APD_{90,M} = 320$ ms, $APD_{90,Epi} = 275$ ms.

In the presence of LQT 2, the low intrinsic density of I_{Ks} caused a disproportionately high prolongation of the APD_{90} in the M cells. However, this prolongation is balanced and thus reduced in tissue simulations due to the effect of electrotonic coupling (e.g. mild LQT 2 (50 % $g_{Kr,max}$) uncoupled: $APD_{90,Endo} = 296$ ms, $APD_{90,M} = 362$ ms, $APD_{90,Epi} = 289$ ms vs. mild LQT 2 (50 % $g_{Kr,max}$) coupled: $APD_{90,Endo} = 349$ ms, $APD_{90,M} = 351$ ms, $APD_{90,Epi} = 296$).

The upper row of Fig. 6.19 shows the ventricular activation sequence for the physiological case. The depolarization was initialized by the stimulus currents at the Purkinje fiber endings in the apical endocardium. In general, the activation sequence was not influenced by the LQT 2 pathology as the associated changes in $g_{Kr,max}$ only affected the phase of ventricular repolarization. No significant potential differences were present during the ST-segment (starting at $t = 77$ ms) resulting in an isoelectric line in the ECG (see Fig. 6.19C). The repolarization sequence in our heterogeneous model started at the endo- and epicardial border of the apex and finally vanished in the basal M cells (see Fig. 6.19D and Fig. 6.19E). This resulted in a concordant T-Wave in the clinical 12-lead ECG (see Fig. 6.20).

In case of LQT 2, the reduction of the repolarizing current I_{Kr} led to a prolongation of the QT interval (see Fig. 6.20). In addition to that, the disproportionately high prolongation of the M cell APD_{90} increased the dispersion of repolarization that was present in the model. An increase of this dispersion led to a widening of the T-

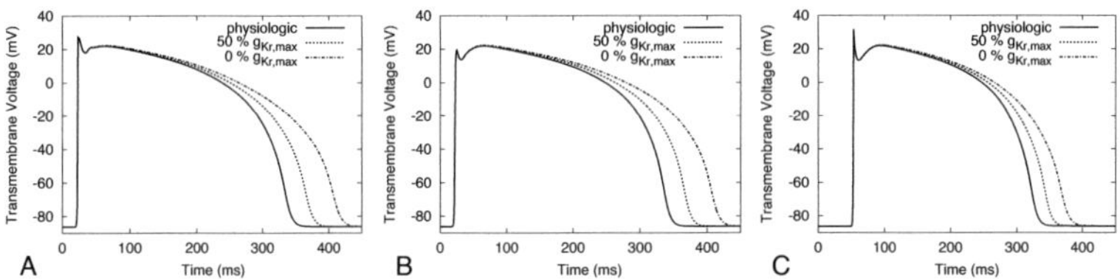

Fig. 6.18. Influence of LQT 2 on AP morphology of endocardial (A), M (B) and epicardial cells (C). The AP traces were extracted from tissue simulations in which the electrotonic coupling smoothened the AP morphology differences between the three cell types. Two degrees of LQT 2 severity were modeled (50 % $g_{Kr,max}$, 0 % $g_{Kr,max}$). The APD_{90} prolongation increased with severity. Figure adopted from [8].

wave and an augmentation of the T-wave amplitude which is visible in Fig. 6.20. Ventricular repolarization was finally finished after $t = 375$ ms, $t = 400$ ms and $t = 433$ ms for the physiological setup, the mild form of LQT 2 (50 % $g_{Kr,max}$) and the more severe form (0 % $g_{Kr,max}$), respectively.

6.4.1.2 Discussion

Although the ECG is still predominantly used in the clinical routine to identify LQTS patients, the quantitative effects of ion channel mutations on the body surface ECG are not entirely understood. Within this context, mutation-induced effects on the T-Wave morphology are of particular interest.

In this study we investigated the effects of LQT 2 on electrotonically coupled ventricular myocytes and the body surface ECG. LQT 2 was modeled either by a reduction of I_{Kr} channel conductivity or a complete block of this ionic current. Due to the heterogeneous electrophysiological properties of the ventricular tissue this led to a disproportionately high prolongation of the APD in the M cells similar to measurement results that were based on a canine ventricular wedge preparation [216]. In addition to that, a QT prolongation was also observed in the computed Einthoven and Wilson leads, as expected.

In spite of the opposite polarity of depolarization and repolarization, our model produced a concordant T-Wave (see Fig. 6.20) which we attributed to the heterogeneous ion channel distribution that was considered. Although it was beyond the scope of this study to evaluate the impact of different heterogenous distributions (e.g. transmural vs. apico-basal) it can be concluded that transmural heterogeneity alone could be sufficient to generate a positive T-Wave. The study from Yan *et al.* [216] supports this hypothesis but nevertheless it should be clarified in future investigations.

It should be noted that although our model was able to reproduce the expected QT prolongation, we did not see a notched or bifurcated T-Wave in any of the simulated standard leads. Yet, such a signal has clinically been associated with an increased risk of Torsade de Pointes [379]. According to Gima and Rudy, T-Wave notches can be induced by superimposing a hypokalemia on the LQT 2 syndrome in a one dimensional model [133]. In this case, the effects of the hypokalemia on the inward rectifier potassium current I_{K1} were held responsible for the appearance of the notch. Although hypokalemia has often been reported to be a potential arrhythmogenic substrate in the presence of LQTS [380] it is not completely clear

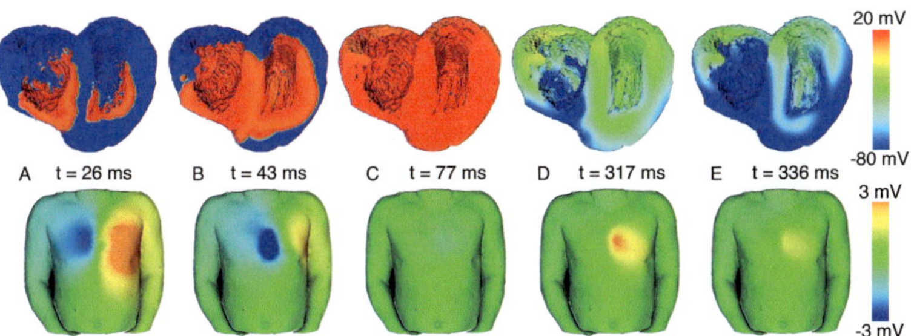

Fig. 6.19. Ventricular activation and repolarization (top row) in the physiological case together with the corresponding BSPMs (bottom row). The depolarization started at $t = 0$ ms in the apical endocardium of the left and right ventricle and propagated first in apical (A) and later in basal epicardial direction (B). After the plateau phase of the AP (ST segment in the ECG) (C), the repolarization sequence started at the apical endo- and epicardium (D) and vanished in the basal midmyocardial region (E). In the displayed physiological case, ventricular repolarization was finished after 375 ms.

if all cases of notched T-Waves can be attributed to hypokalemia or if there might be additional effects involved.

In an extension of the presented model, the ventricular fiber orientation should be considered to allow for the anisotropic conduction velocity that is present in cardiac tissue. Information regarding the course and distribution of ventricular muscle fibers is available from histological investigations [35] and Diffusion Tensor MRI datasets [43]. In addition to that, a computational model of the LQTS would be more realistic if it considered changes in ion channel kinetics that can be extracted from patch-clamp measurements. Finally, additional information on heterogeneous ion channel distributions should be considered. E.g. Volders *et al.* [57] reported a larger I_{Ks} and I_{to} density in the M cells in the right ventricle which abbreviated the APD compared to cells from similar locations in the left ventricle. Furthermore, Szentadrassy et al. [263] found an apico-basal I_{Ks} gradient, which should be implemented in a future version of our model as well.

In general, it could be observed in this study that morphological changes of the T-Wave due to LQT 2 were lead dependent (see Fig. 6.20). In the future it might be possible to use personalized computer-based models of the LQTS to evaluate potential genotype-phenotype correlations. If such correlations could be established, it might be possible to derive special ECG lead positions that amplify these genotype-phenotype relation. In the long run this could lead to an ECG-based genotype identification strategy that avoids the time-consuming and expen-

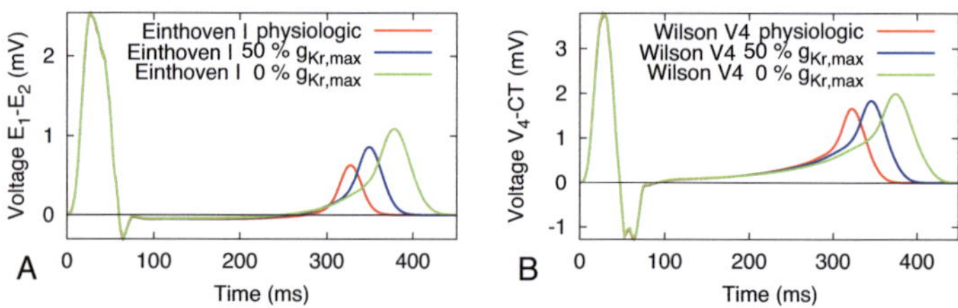

Fig. 6.20. Einthoven I (A) and Wilson V4 (B) ECG leads were extracted from the calculated BSPMs. The increase in dispersion of repolarization due to LQT 2 led to a widening of the T-Wave and an augmentation of the T-Wave amplitude. Figure adopted from [8].

sive DNA analysis, which is up to now still the gold standard for the differentiation between the various types of LQTS.

6.4.2 Results: Suitability of Different Electrophysiological Models to Characterize Long-QT 2

6.4.2.1 Effects of LQT 2 on the APD_{90} and the Transmural ECG

The APD_{90} that was calculated based on the different electrophysiological models under physiological and pathological conditions can be seen in Table 6.2. Without LQT 2, the latest model from ten Tusscher *et al.* (TT 06) [152] showed the largest electrophysiological heterogeneity followed by the model from Seemann *et al.* (SM) [306] and the first version of the model from ten Tusscher *et al.* (TT 04) [151]. In all cases, the introduction of LQT 2 led to a preferential prolongation of the M cell APD_{90} due to the intrinsically weak I_{Ks} current there. The highest absolute and relative M cell APD_{90} prolongation was observed for the SM model while both versions of the ten Tusscher model showed a comparable relative prolongation. Concerning the model from Kurata *et al.* we were surprised to see that the modifications in g_{Ks} and g_{to} did not lead to the desired transmural distribution of APD_{90}. We attributed this to the strong influence which the endocardially reduced I_{to} had on I_{CaL}. Thus the reduction of g_{Ks} in the M cells was not sufficient to prolong the APD_{90} of the M compared to the endocardial cell ($APD_{90,Endo} = 392$ ms vs. $APD_{90,M} = 382$ ms; see Table 6.2).

The effects of LQT 2 on several characteristic signal parameters of the transmural ECG can be seen in Table 6.3. The reduction of the repolarizing I_{Kr} current led to a QT interval prolongation in three of the four models that were evaluated. In addition to that, the associated increase in transmural dispersion of repolariza-

Table 6.2. Comparison between the physiological and pathological APD_{90} in endo, M and epicardial cells.

	Physiological Case			LQT 2 (50% $g_{Kr,max}$)		
	APD_{90} Endo	APD_{90} M	APD_{90} Epi	ΔAPD_{90} Endo	ΔAPD_{90} M	ΔAPD_{90} Epi
TT *et al.* 04 [151]	278 ms	327 ms	271 ms	20 ms 7%	35 ms 11%	19 ms 7%
TT *et al.* 06 [152]	291 ms	413 ms	299 ms	18 ms 6%	43 ms 10%	20 ms 7%
Kurata *et al.* [305]	392 ms	382 ms	341 ms	44 ms 11%	74 ms 19%	46 ms 13%
SM *et al.* [306]	283 ms	338 ms	277 ms	35 ms 12%	82 ms 24%	34 ms 12%

tion resulted in an augmentation of the T-Wave amplitude and a broadening of its morphology.

It was not possible to extract signal parameters from the transmural ECG that was calculated using the Kurata model. The reasons for this were related to the problems that we encountered when we tried to integrate transmural heterogeneity into this model (as described earlier). As no significant T-Wave was present in the corresponding transmural ECG, we excluded the data from the Kurata model from Table 6.3.

Finally, the simulated transmural ECG signals are shown in Fig. 6.21. This allows a visual evaluation of the signal morphology. For comparison, the transmural ECG is shown from a wedge experiment in which LQT 2 was induced pharmacologically [132] (see Fig. 6.21D).

Table 6.3. Comparison of selected features from the transmural ECG under physiological and pathological conditions. Abbreviations: T-Wave full width at half maximum (T_{FWHM}).

	Physiological Case			LQT 2 (50% $g_{Kr,max}$)		
	T_{max}/QRS_{mean}	QT	T_{FWHM}	$\Delta(T_{max}/QRS_{mean})$	Δ QT	Δ T_{FWHM}
TT *et al.* 04 [151]	0.094	335 ms	31 ms	0.068 72%	28 ms 8%	3 ms 10%
TT *et al.* 06 [152]	0.254	386 ms	37 ms	0.047 18%	40 ms 10%	12 ms 32%
SM *et al.* [306]	0.167	328 ms	15 ms	0.118 70%	63 ms 19%	9 ms 60%

6.4.2.2 Discussion

In general, characteristic features of LQT 2 that were found in wedge experiments [132] (like QT interval prolongation and increase in transmural dispersion of repolarization that led to larger and wider T-Waves) were reproduced correctly by both models from ten Tusscher *et al.* [151, 152] and by the model from Seemann *et al.* [306]. Differences were only observed regarding the degree of existing electrophysiological heterogeneities as well as with respect to the changes that were introduced by LQT 2. However, it should be noted that the ventricular wedge preparation, which we used as reference for this study, was also only a model (based on pharmacological agents) of LQT 2. The large amplitude T-Waves that were seen after the onset of LQT 2 are in contrast to clinically recorded body surface ECG signals which show mainly low amplitude T-Waves for patients suffering from LQT 2 [119].

In a prospective study, the various electrophysiological models including the LQT 2 modifications could be used to simulate the body surface ECG. The results could then be compared to patient data and thus allow a more realistic assessment of each model's performance.

6.4.3　Results: In-silico Evaluation of Beta-Adrenergic Effects on the Long-QT Syndrome

6.4.3.1　Transmural ECG and Transmural Dispersion of Repolarization

After the wedge was electrically activated at the endocardium, the excitation wave propagated towards the epicardium. Due to the chosen APD_{90}-distribution (short APD_{90} in epicardial vs. long APD_{90} in M cells, see Fig. 4.10A), the repolarization started in the epicardial region and ended in the M cells. The transmurally adapted tissue conductivity arrangement (see Fig. 4.10B) resulted in a conduction time of 26 ms. Based on the geometrical dimensions of the wedge, this conduction time could be translated into a conduction velocity of 46 cm/s, which was close to experimental recordings [262].

The AP plots for all LQT mutations (Fig. 6.22-Fig. 6.24) were extracted from the tissue simulations in which electrotonic coupling was present. The APD_{90} changes reported in the following are relative changes with respect to the wild-type (WT) setup.

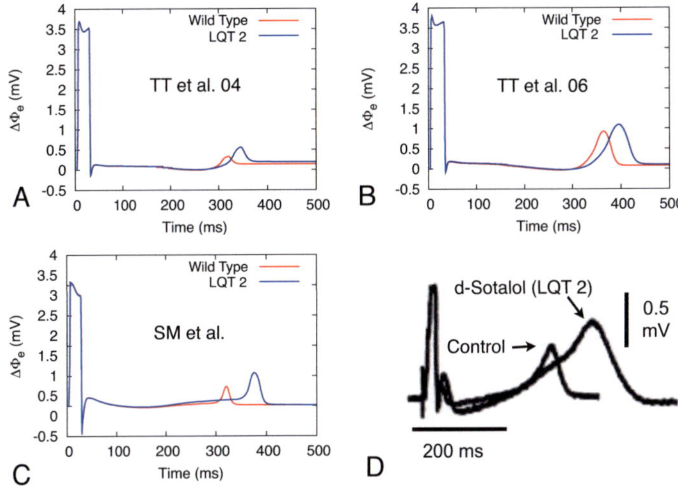

Fig. 6.21. Overview over the signal morphology of the transmural ECGs that were calculated using different electrophysiological models: A: The transmural ECG from the first version of the ten Tusscher model [151] had a small T-Wave and showed only moderate QT prolongation after the onset of LQT 2. B: In contrast to that, the revised model from ten Tusscher *et al.* [152] showed a larger T-Wave but similar QT prolongation. C: Finally the model from Seemann *et al.* [306] showed the largest LQT 2 related changes. In this case, the small T-Wave (under physiological conditions) became substantially larger and wider and the QT time was prolonged significantly (after LQT 2 was induced). D: Transmural ECG resulting from a wedge experiment in which LQT 2 was induced pharmacologically (based on d-Sotalol) [132]. Figure modified based on [132].

LQT 1

The effects of a loss of function of I_{Ks}, which was used to model LQT 1, can be seen in Fig. 6.22. After reducing $g_{Ks,max}$ by 50%, the APD_{90} of M and epicardial cells were both prolonged ($APD_{90,M}$ +16%, $APD_{90,EPI}$ +21%). If we additionally activated the adrenergic effects by virtually administering $1\mu M$ ISO, the APD_{90} of the M cells was slightly longer than in LQT1 without ISO ($APD_{90,M}$ +21%) whereas the APD_{90} of epicardial cells was abbreviated by 4% (see Fig. 6.22A, Fig. 6.22B). With respect to the TDR, we saw that a simple block of $g_{Ks,max}$ (with varying degree) did not lead to significant changes. In contrast to that, the TDR rose substantially after we applied ISO in addition to the reduction of $g_{Ks,max}$ (see Fig. 6.22D). Finally, the main effect of LQT 1 on the tECG was a QT prolongation whereas the T-Wave morphology was not affected. The application of ISO compensated this QT prolongation. In addition to that, it was surprising to see that the increase of TDR under ISO influence did not lead to a widening of the T-Wave or an increase of its amplitude (see Fig. 6.22G).

LQT 2

LQT 2 was modeled by a reduction of $g_{Kr,max}$. This led to a slight prolongation of the APD_{90} in both M and epicardial cells ($APD_{90,M}$ +7%, $APD_{90,EPI}$ +6%, see Fig. 6.23A and Fig. 6.23B). In this case, the activation of the adrenergic signaling cascade caused a dramatic shortening of epicardial APD_{90} (-21%) whereas M cells were not affected (+7%) to the same extent. Unlike seen in LQT 1, an increase in the block of the affected channel (now: I_{Kr}) did lead to an increase in TDR here (see Fig. 6.23D). This effect was even enhanced by the application of ISO. With respect to the tECG, we observed a slight QT prolongation and widening of the T-Wave in case of active LQT 2 (see Fig. 6.23F). However, no additional widening was seen after the application of ISO (the T-Wave even became narrower) even though the TDR was largest in this case.

LQT 3

After the mutation 1795insD was included in I_{Na} to model LQT3 both the APD_{90} of M and epicardial cells were marginally prolonged (+5% and +3%, respectively). The addition of ISO had the largest effects on the APD_{90} of epicardial cells ($APD_{90,M}$ +2%, $APD_{90,EPI}$ -23%, see Fig. 6.24A and Fig. 6.24B) thereby increasing the TDR ($TDR_{WT} = 47ms$, $TDR_{LQT3} = 53ms$, $TDR_{LQT3,ISO} = 96ms$, see Fig. 6.24D). The effects of the mutation on the tECG were rather small (see Fig. 6.24F). A slight QT prolongation could be observed after the mutation was activated while the application of ISO shortened the QT interval to even smaller values than in the WT case.

6.4.3.2 Discussion

The aim of this study was the reproduction of the wedge experiments, in which Antzelevitch *et al.* analyzed the effects of LQT 1-3 under the influence of beta-adrenergic regulation. Previous in-silico models [227, 133] were limited by the fact that no model was available at that time to describe the intracellular adrenergic signaling cascade.

The first step towards the reproduction of the experiments was to create similar experimental conditions: To this end, we adapted $g_{Ks,max}$ in the endocardial, M

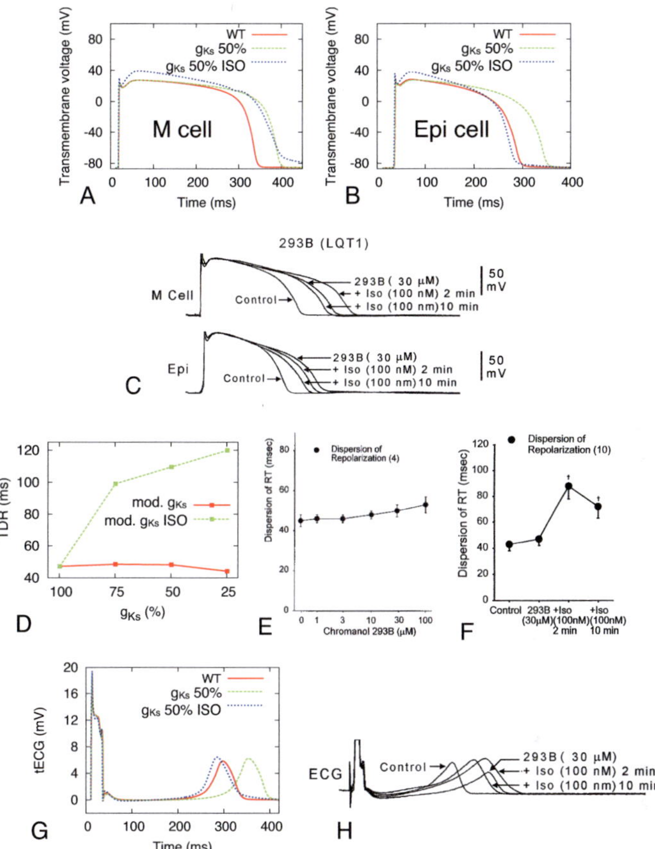

Fig. 6.22. AP morphology, development of TDR and transmural ECG for a model of LQT 1. A,B: The AP morphology of M (A) and epicardial (B) cells is shown for the WT case, LQT 1 (50% $g_{Ks,max}$) and LQT 1 with ISO. C: The AP morphology from canine wedge experiments is shown for comparison (Figure directly adopted from [221]). At least a qualitative agreement can be seen: while the $APD_{90,M}$ remains unchanged (or is even slightly prolonged) in case of M cells and active LQT 1 + ISO, the epicardial $APD_{90,EPI}$ is shortened after the application of ISO. D: Development of the TDR for different degrees of channel block with and without ISO. The TDR was almost unaffected by a simple channel block, while the addition of ISO increased it significantly. E: Experimental data showing the dispersion of repolarization times (which is not identical, but related to the TDR) for different concentrations of the channel blocker Chromanol 293B (Figure directly adopted from [217]). F: Experimental data showing again the dispersion of repolarization times in the presence of the channel blocker Chromanol 293B while ISO was administered additionally during the measurements (Figure directly adopted from [221]). G: Simulated transmural ECG. H: Measured transmural ECG (Figure directly adopted from [221]).

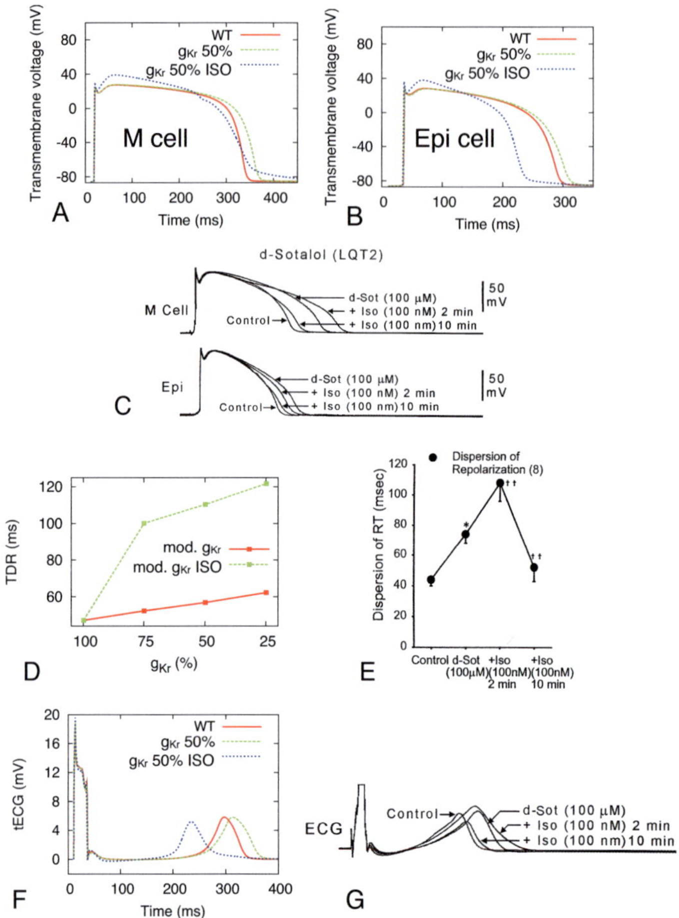

Fig. 6.23. AP morphology, development of TDR and transmural ECG for a model of LQT 2. A,B: The AP morphology of M (A) and epicardial (B) cells is shown for the WT case, LQT 2 (50% $g_{Kr,max}$) and LQT 2 with ISO. C: The AP morphology from canine wedge experiments is shown for comparison (Figure directly adopted from [221]). D: Development of the TDR for different degrees of channel block with and without ISO. The TDR did slightly increase with an increase in the degree of the channel block. If ISO was administered additionally the TDR increased significantly. E: Experimental data showing the dispersion of repolarization times (which is not identical, but related to the TDR) in the presence of the channel blocker d-Sotalol and ISO (Figure directly adopted from [221]). F: Simulated transmural ECG. G: Measured transmural ECG (Figure directly adopted from [221]).

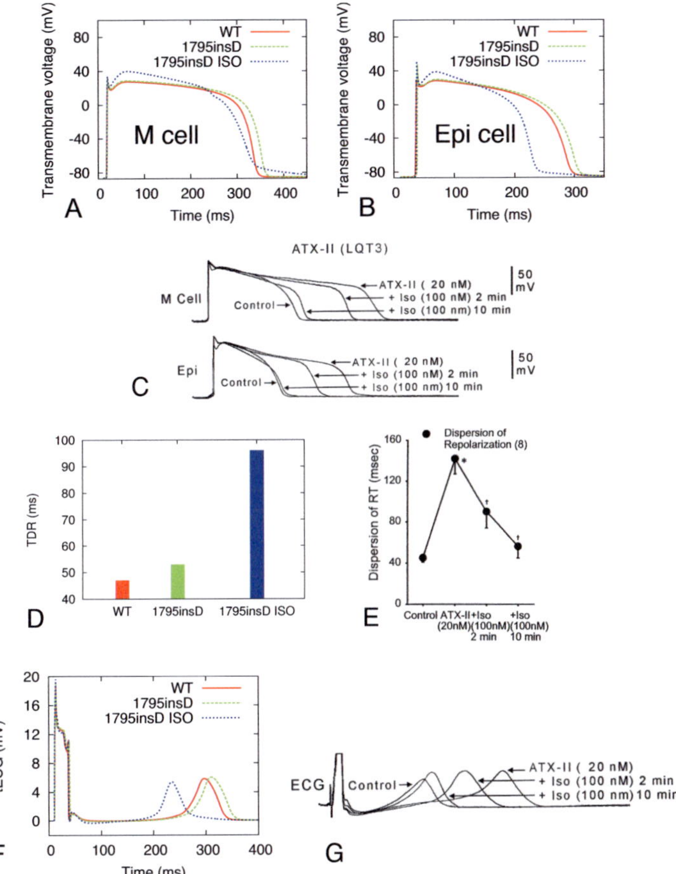

Fig. 6.24. AP morphology, development of TDR and transmural ECG for a model of LQT 3. A,B: The AP morphology of M (A) and epicardial (B) cells is shown for the WT case, LQT 3 (mutation 1795insD) and LQT 3 with ISO. C: The AP morphology from canine wedge experiments is shown for comparison (Figure directly adopted from [221]). D: Development of the TDR in case of active LQT 3 with and without ISO. The TDR did slightly increase in case of active LQT 3. If ISO was administered additionally the TDR increased significantly. E: Experimental data showing the dispersion of repolarization times (which is not identical, but related to the TDR) in the presence of ATX-II (which was used to model LQT 3) and ISO (Figure directly adopted from [221]). It was interesting to see, that ISO reduced the TDR in the experiment. This did not agree with the model or our theoretical considerations. F: Simulated transmural ECG. G: Measured transmural ECG (Figure directly adopted from [221]).

and epicardial layer in order to mimic the transmural APD_{90}-distribution that was measured in the wedge (see Fig. 4.10A). As the exact position and thicknesses of the three transmural layers were not explicitly stated, we had to assume a certain distribution (in this case: 20%:30%:50%). However, with this setup we were not able to completely reproduce the transmural course of the APD_{90} even though we added a layer of high resistivity that was reported to decouple the epicardium from the M cells [119] (see Fig. 4.10B). The existence of this layer of high resistivity was attributed to a region with a sharp transition in fiber orientation [119]. Yet such a sharp transition in fiber orientation has never been reported by any other study (see section 2.1.2). More recent studies from Glukhov *et al.* [155] and Poelzing *et al.* [340, 341] attributed this increase in resistivity to a reduced density of connexin 43 in epicardial cells (connexin 43 is an indicator for gap junction density).

The results of the computational wedge experiments were in good qualitative (but not quantitative) agreement with the measurements. Concerning LQT 1, the reduction of $g_{Ks,max}$ induced a homogeneous prolongation of the APD_{90} in all cell types without a significant reduction in TDR [217]. Only a large reduction to 25% of the baseline $g_{Ks,max}$ value led to a slight reduction in TDR (see Fig. 6.24D). This is different from findings in previous studies [227, 226] where a reduction of the heterogeneously distributed I_{Ks} always led to a reduced dispersion of repolarization and thereby to narrow or even negative T-Waves. The main reason for the almost constant TDR that we observed in this study was seen in the initial elevation of baseline $g_{Ks,max}$ values, which were adapted in order to reproduce the measured course of APD_{90} (see Fig. 4.10A). In this case, we assumed that the heterogeneously distributed current I_{Ks} was expressed with a significantly higher density than the homogeneously distributed current I_{Kr}. This means that we could moderately reduce $g_{Ks,max}$ while still preserving the existing dispersion of repolarization. This effect can also be seen in the study from Gima *et al.* [133] in which extremely high densities of I_{Ks} were chosen compared to I_{Kr} (11:1 endocardial, 4:1 midmyocardial and 35:1 epicardial). Based on this parameterization, they were able to generate positive T-Waves (in case of a partial I_{Ks} block) with similar width and amplitude as in the WT case. Yet they also claimed upright T-Waves even if I_{Ks} was set to 0%. This is quite surprising as they reported in the same study that an electrophysiologically homogeneous wedge preparation will produce negative T-Waves.

Concerning LQT 2, a reduction of the homogeneously distributed I_{Kr} amplified the intrinsic I_{Ks} heterogeneity which in turn led to an increase in TDR and to a widening of the T-Wave in the tECG. The same was true for LQT 3 where the reactivation of I_{Na} caused a preferential prolongation of the APD_{90} in M cells and thus increased the TDR. The models of both LQT 2 and LQT 3 led to a much smaller APD_{90} (on cell level) and QT (concerning the tECG) prolongation compared to LQT 1. This was due to the elevated levels of I_{Ks}: in case of I_{Kr} the density was low compared to I_{Ks}, therefore a reduction did not lead to a significant prolongation. In case of the mutated I_{Na}, the reactivation was not enough to significantly prolong the APD_{90}.

In all three mutations, the application of ISO always shortened the APD_{90} of epicardial cells. In case of the M cell, ISO prolonged the APD_{90} in the presence of LQT 1 and LQT 2 while there was almost no ISO-induced change in the presence of LQT 3. In addition to that, the TDR was increased in the presence of ISO (for all three mutations), which was in agreement with the experimental findings in case of LQT 1 and LQT 2 but in contradiction with reports from LQT 3 where a TDR reduction has been reported [221]. In our simulations, however, an ISO-induced increase in TDR did not lead to a significant widening of the T-Wave or to an increase of its amplitude. We attributed this effect to a combination of the AP morphology changes in the presence of ISO (flat slope at the repolarizing flank) and the specific features of the APD_{90}, which is extracted exactly in the region of this flat slope. In this case, this special AP morphology leads to large APD_{90} values but at the same time, the potential difference that is present in the computational model of the wedge is small. For more information on this ISO-induced AP morphology changes please refer to section 6.3.3. If we consider the ISO effects over time, we were surprised that our computational results matched with the effects that Antzelevitch *et al.* observed 2 minutes after the application of ISO. This was unexpected, as our pre-calculations ensured that the ISO-effects were maximal in the in-silico model. In theory, these maximal effects should rather correspond to the effects that Antzelevitch *et al.* observed after 10 minutes in their experiment.

In this study, we were able to reproduce some of the features of the pharmacologically induced LQT syndromes from the well-known wedge experiments. However, it should be noted that these experiments were based on modeled (pharmacologically) rather than inherited LQT pathologies. When compared to clinical recordings from LQT patients there are some obvious contradictions:

- Although the wedge experiments predict no significant changes in TDR and T-Wave morphology for LQT 1, patients suffering from this disease exhibit broad-based T-Waves even at rest and without adrenergic regulation [118]. Yet, the appearance of these broad-based T-Waves is difficult to explain: Under physiological conditions, the heterogeneous I_{Ks} distribution is generally assumed to be responsible for the dispersion of repolarization in ventricular tissue. Now, a loss of function mutation affecting I_{Ks} (like LQT 1) should reduce the dispersion, which would lead to narrow T-Waves of low amplitude rather than broad-based T-Waves.

- Concerning LQT 2, the situation is vice versa: In this case, the wedge experiments predict an increase in TDR as the mutation reduces the conductivity of I_{Kr}, thereby preferentially prolonging the APD_{90} in M cells (due to the low intrinsic density of the other repolarizing current I_{Ks}). This directly contradicts the low-amplitude T-Waves, which are clinically observed in LQT 2 patients [118].

One possible explanation for this obvious contradiction in case of LQT 1 might be found aside the classical loss of function mutation. Recently, a mutation has been identified that does not reduce the conductivity of I_{Ks} but rather damages a protein that is responsible for its link to the adrenergic signaling cascade [100]. As all other channels are still modulated by the sympathetic nervous system, the application of ISO (or physical activity) will cause an increased inflow of e.g. Ca^{2+} (as I_{CaL} is enhanced by ISO), which will prolong the APD_{90}. Under normal conditions, the ISO-induced increase in I_{Ks} will restrict the APD_{90} prolongation, which is not possible in case of the reported mutation (as the adrenergic regulation of I_{Ks} was deactivated by the mutation).

7

Results: Anatomical Modeling and the Forward Problem of Electrocardiography

This chapter presents the results of the various studies in the realm of *anatomical modeling and the solution of the forward problem* that were conducted during the course of this thesis. The methods that were used during the creation of these results are described in section 5.

7.1 Results: Creating Anatomical Models based on MRI data

7.1.1 Overview: Clinically Acquired Data

Table 7.1 provides an overview over the data that was acquired in cooperation with the University Hospital Heidelberg. In total, multi-channel ECG data were recorded for 6 patients/probands, while MRI scans were performed on 4 of the 6 patients/probands. The male and female probands were healthy while all patients suffered from the congenital long QT syndrome. For 2 patients, no long QT subtype was known while for the remaining 2 patients the subtype was known (LQT1) yet the exact mutations still had to be identified.

MR imaging was performed on a 1.5 T Magnetom Avanto scanner (Siemens Medical Systems, Erlangen, Germany). The torso was always imaged both in inspiratory and expiratory breath-hold (resolution: $\approx$1x1x2 mm^3). The heart was imaged in diastolic state ($\approx$1x1x1 mm^3). In one case, we additionally acquired 4D cinematographic and tagging data of the ventricles, which we used to build a model of the ventricular contraction and relaxation (see section 5.4).

Multi-channel ECG data were acquired using a 64 or 80 lead-system (ActiveTwo, BioSemi, Amsterdam, Netherlands). In order to be able to use the same electrode positions for the extraction of the simulated ECGs that were also used during the ECG measurements, the electrode positions were localized with an electromagnetic tracking system (FASTRAK, Polhemus, Burlington VT, USA), see section 2.3.

7.1.2 Visualization of Segmented and Classified Heart and Torso Models

Fig. 7.1 - 7.4 show the available MRI raw data, the segmented and classified voxel models that were created based on the raw data and finally 3D visualizations of the resulting heart and torso models. Different MRI sequences were used to image torso and heart. Both datasets were independently segmented and subsequently combined for the 3D torso model. Only coarse atrial segmentation was performed in this work as the atria were only included in the torso model to establish a realistic surrounding for the ventricles during the solution of the forward problem. The detailed atrial structures that are visible in the 3D heart model were segmented by Krueger *et al.* [314, 315, 316].

7.2 Results: Resolution Effects of Anatomical Models

7.2.1 Investigations on Wavefront Shape

The isochrone plots in Fig. 7.5 visualize the excitation spread in the three virtual tissue patches (isochrone rings correspond to the wavefront position in time increments of 10 ms). It is obvious that the shape of the wavefront was different for each resolution. The largest differences were observed close to the angel bisector of the major and minor semi-axes x and y (marked in Fig. 7.5). In theory, the wavefront should have the shape of an ellipsoid (radii of the ellipsoid depend on the anisotropy ratio). In order to evaluate the difference between this elliptical reference and the different resolution-dependent excitation wavefronts quantitatively we determined the maximum difference between the fourth isochrone ring (see arrow in Fig. 7.5) and a manually fitted ellipsoid. As expected, the differences were larger for lower resolution. For a resolution of 0.4 mm the maximum difference was 1.13 mm while it was reduced to 0.56 mm and 0.22 mm for resolutions of 0.2 mm and 0.1 mm, respectively. Due to these results and previous reports in the literature [166], we chose to conduct the simulations in the biventricular model with a voxel size of 0.2 mm. This resolution provides a good compromise between realistic excitation patterns and computational complexity.

Table 7.1. Overview over the clinically acquired data from patients and probands. MRI imaging was performed on a 1.5 T Magnetom Avanto scanner (Siemens Medical Systems, Erlangen, Germany). Multi-channel ECG data was acquired using a 64 or 80 lead-system (ActiveTwo, BioSemi, Amsterdam, Netherlands). Electrode positions were located with an electromagnetic tracking system (FASTRAK, Polhemus, Burlington VT, USA). † For this dataset we acquired 4D cinematographic and tagging data of the ventricles in addition to the static diastolic scan (see section 5.4). ‡ This patient suffered from syncopes and had an implantable cardioverter-defibrillator (ICD). In all cases, the torso was imaged in inspiratory and expiratory breath-hold. Usually, the expiratory dataset was used for the construction of the torso model (exception: dataset marked with ∗). F: Female, M: Male, N/A: not available

ID	Sex	Size	Weight	Age	Acquisition Date	Pathology	Torso MRI	Heart MRI	ECG	FASTRAK
1	F	1.60 m	52 kg	47	15/08/2007	No	$\approx$1x1x2 mm^3	$\approx$1x1x1 mm^3	64 channel	N/A
2	F	1.63 m	66 kg	18	15/08/2007	LQT (Subtype N/A)	$\approx$1x1x2 mm^3	$\approx$1x1x1 mm^3	64 channel	N/A
† 3	M	1.80 m	79 kg	27	19/03/2008	No	$\approx$1x1x2 mm^3	1x1x1 mm^3	64 channel	Yes
4	F	N/A	N/A	17	19/02/2010	LQT (Subtype N/A)	N/A	N/A	80 channel	N/A
‡ 5	F	N/A	N/A	48	16/03/2010	LQT1 (Mutation S2772)	N/A	N/A	80 channel	N/A
∗ 6	F	1.70 m	79 kg	50	14/05/2010	LQT1 (Mutation S2772)	$\approx$1x1x2 mm^3	1x1x1 mm^3	80 channel	N/A

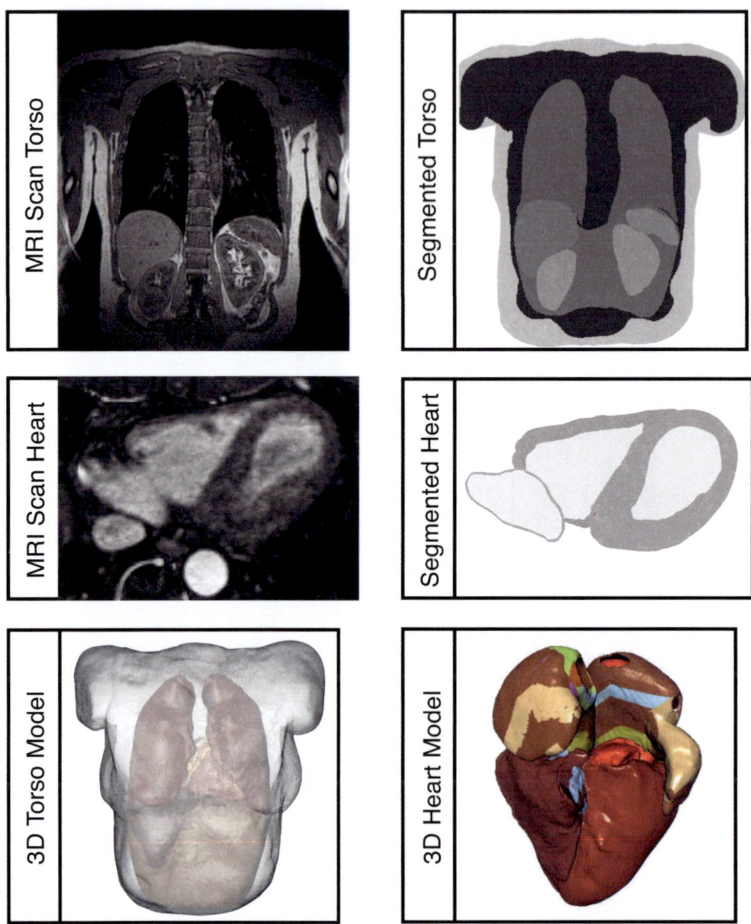

Fig. 7.1. Raw MRI data, segmented voxelized models and 3D visualizations of the data with patient/proband-ID 1 (see Table 7.1). Different MRI sequences were used to image torso and heart (the heart dataset also had a higher resolution). Both datasets were independently segmented and subsequently combined for the 3D torso model shown in the bottom part of the figure. Only coarse atrial segmentation was performed in this work as the atria were only included in the torso model to establish a realistic surrounding for the ventricles during the solution of the forward problem. The detailed atrial structures that are visible in the 3D heart model were segmented by Krueger *et al.* [314, 315, 316].

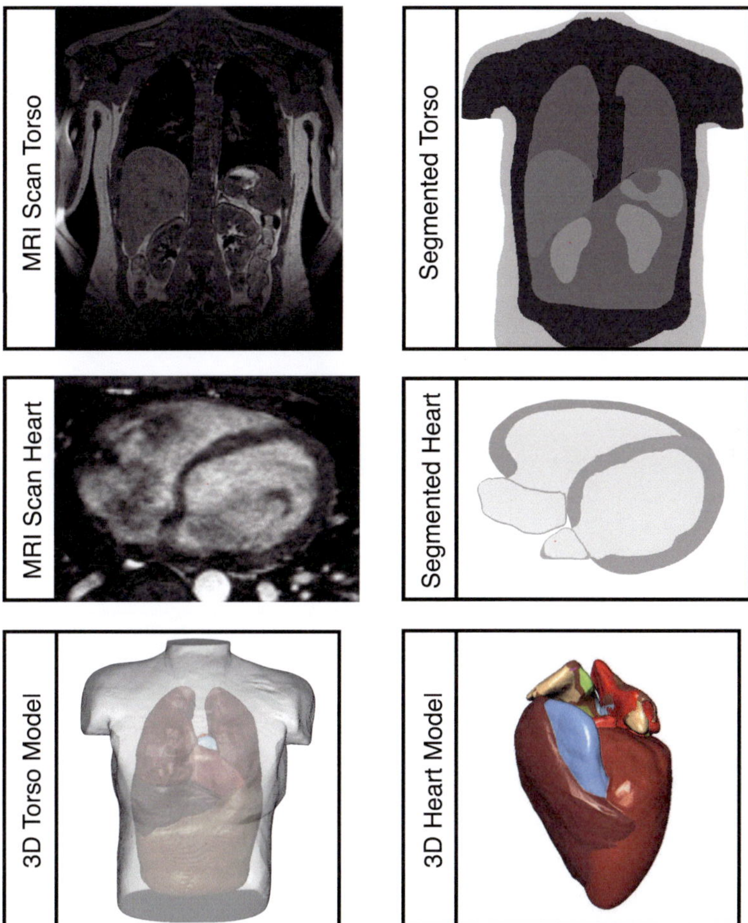

Fig. 7.2. Raw MRI data, segmented voxelized models and 3D visualizations of the data with patient/proband-ID 2 (see Table 7.1). Different MRI sequences were used to image torso and heart (the heart dataset also had a higher resolution). Both datasets were independently segmented and subsequently combined for the 3D torso model shown in the bottom part of the figure. Only coarse atrial segmentation was performed in this work as the atria were only included in the torso model to establish a realistic surrounding for the ventricles during the solution of the forward problem. The detailed atrial structures that are visible in the 3D heart model were segmented by Krueger *et al.* [314, 315, 316].

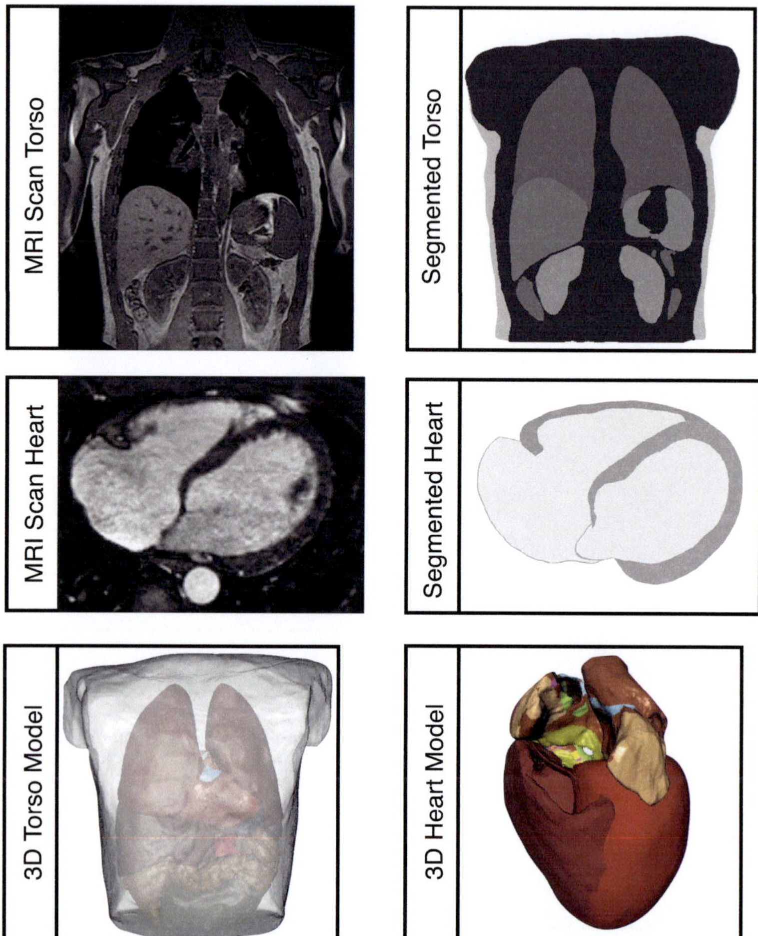

Fig. 7.3. Raw MRI data, segmented voxelized models and 3D visualizations of the data with patient/proband-ID 3 (see Table 7.1). Different MRI sequences were used to image torso and heart (the heart dataset also had a higher resolution). Both datasets were independently segmented and subsequently combined for the 3D torso model shown in the bottom part of the figure. Only coarse atrial segmentation was performed in this work as the atria were only included in the torso model to establish a realistic surrounding for the ventricles during the solution of the forward problem. The detailed atrial structures that are visible in the 3D heart model were segmented by Krueger *et al.* [314, 315, 316].

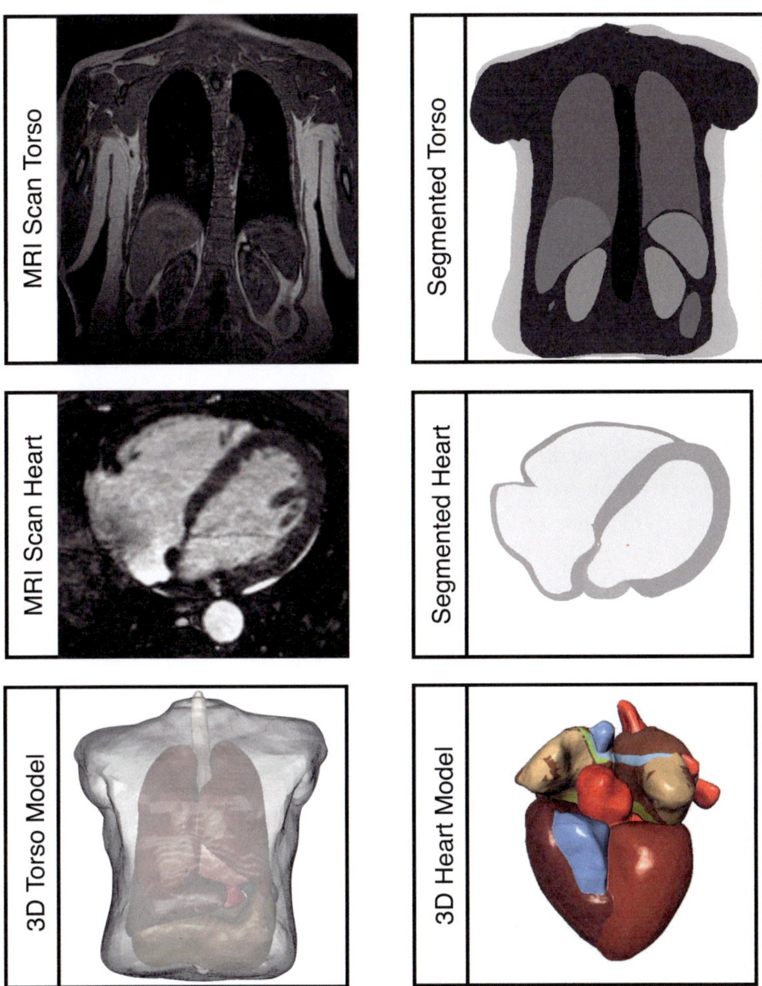

Fig. 7.4. Raw MRI data, segmented voxelized models and 3D visualizations of the data with patient/proband-ID 6 (see Table 7.1). Different MRI sequences were used to image torso and heart (the heart dataset also had a higher resolution). Both datasets were independently segmented and subsequently combined for the 3D torso model show in the bottom part of the figure. Only coarse atrial segmentation was performed in this work as the atria were only included in the torso model to establish a realistic surrounding for the ventricles during the solution of the forward problem. The detailed atrial structures that are visible in the 3D heart model were segmented by Krueger *et al.* [314, 315, 316].

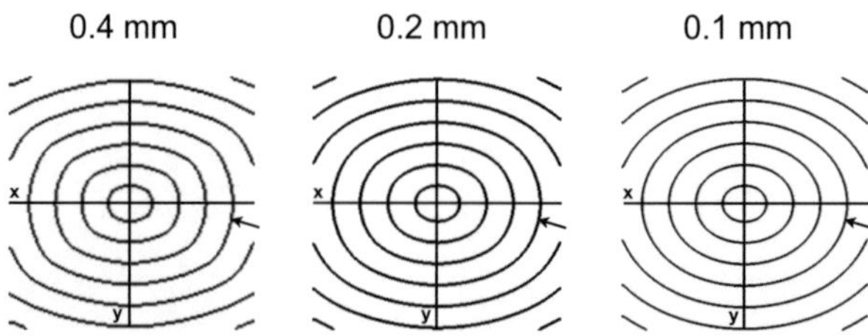

Fig. 7.5. Isochrones of the action potential propagation in three patches of ventricular tissue with different resolutions (isochrone rings correspond to the wavefront position in time increments of 10 ms). The major and minor semi-axes are labeled with x and y, respectively. Arrows mark the fourth isochrone ring. This ring was used for quantitative evaluation of the wavefront shape.

7.2.2 Biventricular Model

Fig. 7.6 shows a visualization of the excitation spread in the biventricular model. Activation was initiated at the sites of early endocardial activation as reported by Durrer *et al.* [174]. Intracellular conductivities were set to $\sigma_{i,T} = 0.072$ S/m (transverse) and $\sigma_{i,L} = 0.114$ S/m (longitudinal). The resulting conduction velocities were 50 cm/s in transverse and 65 cm/s in longitudinal direction.

The simulation was performed on 5 Apple Xserves. Each Xserve had 2 quad core processors (2.8 GHz Intel Xeon) and at least 24 GB of memory. The simulation of the excitation spread (100 ms of cardiac activity) took approximately 5 hours and occupied 100 GB of memory.

7.2.3 Discussion

In this study, we investigated the effects of spatial resolution on the wavefront shape and presented a highly detailed biventricular model that allowed the realistic simulation of ventricular activation. In-silico experiments on virtual tissue patches with different resolutions demonstrated that lower resolutions led to a deformation of the excitation wavefront especially close to the angel bisector of the major and minor semi-axes that characterizes the elliptical shape of the ideal-theoretic case. In a previous study, it had already been shown that the shape of the wavefront depends on the spatial resolution of the model [230]. However in that study, these changes were partially due to the associated changes in conduction velocity as the intracellular conductivities were kept constant. In the experiments presented in this study, we adapted the intracellular conductivities in order to preserve the lon-

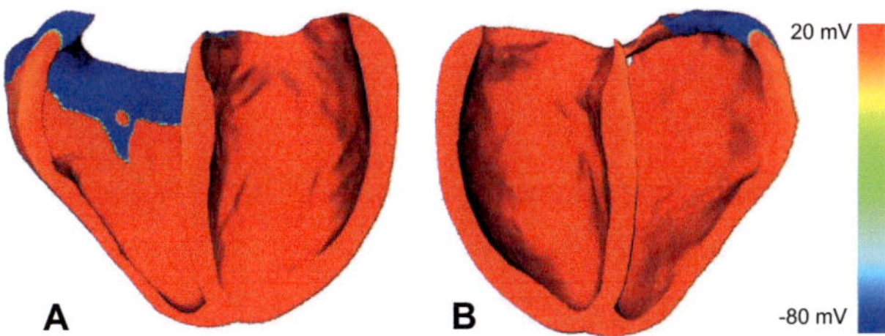

Fig. 7.6. Visualization of ventricular activation in the biventricular model after 85 ms. Stimulus currents were applied according to a complex stimulation protocol in an effort to mimick the specialized excitation conduction system. A: Anterior view. B: Posterior view.

gitudinal and transverse conduction velocities. But even with identical conduction velocities, the excitation fronts showed resolution-dependent differences. These differences were attributed to changes in coupling between the individual computational cells and different neighborhood relations due to stair-case artifacts at lower resolutions.

In case of the biventricular model, we would like to emphasize that it was the largest model that has been used to simulate cardiac electrophysiology at the Institute of Biomedical Engineering (Karlsruhe Institute of Technology) so far. In addition to that, it is to the best of our knowledge the first model that incorporates the following features:

- High spatial resolution (0.2 mm)
- Realistic fiber orientation
- Transmural and apico-basal electrophysiological heterogeneities
- Rule-based stimulation profile that enabled a physiological excitation sequence

The electrophysiological simulations were conducted based on a voxelized representation of the ventricular anatomy. Such a structured grid had the disadvantage that the majority (84.5%) of the elements did not contain excitable tissue. As structured grids do not allow to discretize these areas with a lower resolution, a large amount of memory overhead is created in each simulation. To avoid this problem, we envision the use of unstructured tetrahedron meshes to represent the cardiac geometry during electrophysiological simulations in the future.

In conclusion it can be said that the presented model lay the foundations for future studies in which we want to couple highly detailed biventricular models to a computational representation of the thorax. This will enable us to predict the effects of changes on the ion-channel level up to the body surface ECG (e.g. see section 4.2).

7.3 Results: Modeling Skeletal Muscle Fiber Orientation and the Effects on the ECG

7.3.1 Models of Skeletal Muscle Fiber Orientation

A visualization of the skeletal muscle fiber orientation models is shown in Fig. 7.7. When compared with the simplified rule-based models (Fig. 7.7C-E), the gold standard (Fig 7.7A-B) showed a much more complex fiber arrangement. Differences were particularly large in the abdominal region, as well as near the arms, the neck and the back muscles as all fibers in these regions had a significant longitudinal component. All rule-based setups (exception: Gradient+Back) neglected possible longitudinal fiber orientation components. Please note, that the method proposed by Klepfer *et al.* (Fig. 7.7C) showed discrete transitions every $30°$ degrees which are marked by small arrows in Fig. 7.7C. These discrete transitions were due to the discrete bisectors of the twelve cross-section segments (see Fig. 5.5). In contrast to that, the proposed gradient approach (Fig. 7.7D) created a smoothened circumferential orientation. The only rule-based setup that partially considered longitudinal fiber orientation components was the setup Gradient+Back. However, even in this case, longitudinal components were limited to the back muscle region and transitional orientations were not considered.

Table 7.2. RMSE (μV) between gold standard and rule-based fiber orientation setups

Fiber Orientation Setup	RMSE
Klepfer	88
Gradient	89
Gold+No-Z	83
Gradient+Back	55
Only-Heart	40

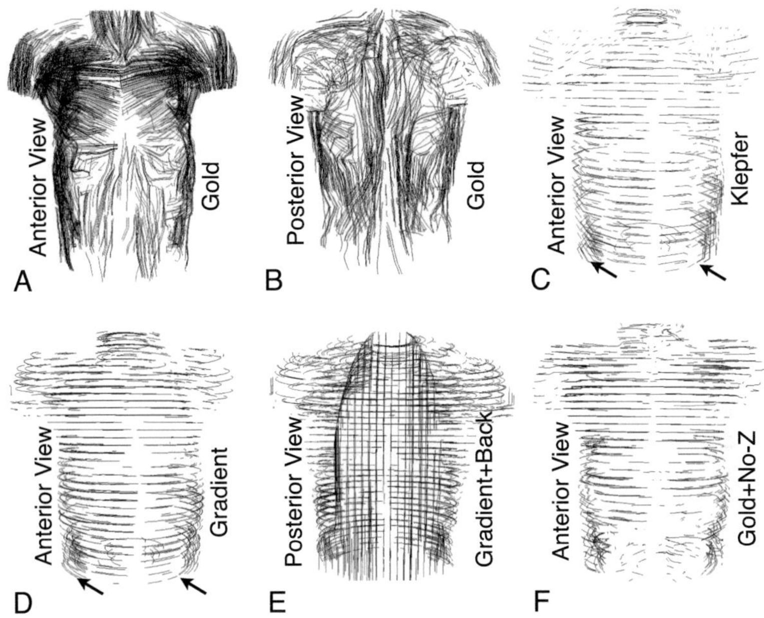

Fig. 7.7. Visualizations of the different fiber orientation setups evaluated in this study. The visualizations were constructed by following the fiber orientation starting from user-defined seed points. A,B: Complex fiber orientation of the gold standard (Gold). It can be seen, that muscles in the abdominal region, the arms, neck and back have significant longitudinal components. C: Rule-based fiber orientation setup according to Klepfer *et al.* [243] (Klepfer). The arrows mark sharp fiber orientation transitions. D: Rule-based fiber orientation setup according to the new gradient approach (Gradient). In this case, the sharp fiber orientation transitions that were visible in Klepfer model could be avoided. E: Rule-based fiber orientation setup based on the gradient approach. Here, the longitudinally oriented back muscles were considered as well. F: Rule-based fiber orientation setup based on the gold standard. However, in this case, the longitudinal fiber vector components were set to zero.

7.3.2 Effects on the ECG and BSPM

Fig. 7.8 and Fig. 7.9 show the impact of the different fiber orientation setups on the Einthoven I and Einthoven II leads, respectively. The upper part of each figure shows the corresponding ECG of the Gold setup which we used as reference. It is interesting to note that the changes in the QRS complex strongly depended on the ECG lead. When comparing Fig. 7.8 with Fig. 7.9, amplitude changes were relatively large in the Einthoven I lead while they were much smaller in Einthoven II. During ventricular repolarization (T-Wave) signal changes could be observed

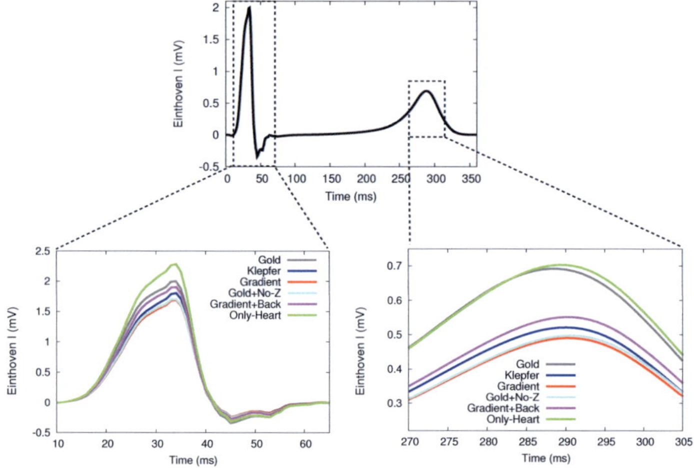

Fig. 7.8. The upper part of the figure shows the Einthoven I ECG calculated with the gold standard fiber orientation. The lower part shows a magnification of the QRS complex and the T-Wave allowing a visual evaluation of the performance of the simplified setups. It can be seen, that all simplified setups (except Only-Heart) underestimated the QRS complex and the T-Wave amplitudes.

in both leads. In this case, however, the signal amplitude was underestimated by the simplified setups in Einthoven I but overestimated in Einthoven II.

If the evaluation was focused on the first two Einthoven leads, no coherent conclusion could be drawn as two setups performed equally well:

- Gradient+Back: this setup showed the best match with the reference concerning the QRS complex of the Einthoven I lead and the T-Wave in Einthoven II
- Only-Heart: this setup performed best with respect to the QRS complex of the Einthoven II lead and the T-Wave in Einthoven I

The results of a more universal evaluation of the BSPM changes (which is not limited to the first two Einthoven leads) are summarized in Table 7.2. In this case, the RMSE was calculated between the Gold setup and the various simplified setups over all 64 electrodes (over all time-steps). All rule-based approaches that neglected the longitudinal fiber orientation components showed similar RMSE values (e.g. compare Klepfer and Gradient). A comparable RMSE was also observed, if the longitudinal components in the gold standard setup were set to zero (Gold+No-Z). If the longitudinal orientation of the back muscles was considered in the rule-based approach (Gradient+Back) the RMSE showed significant improve-

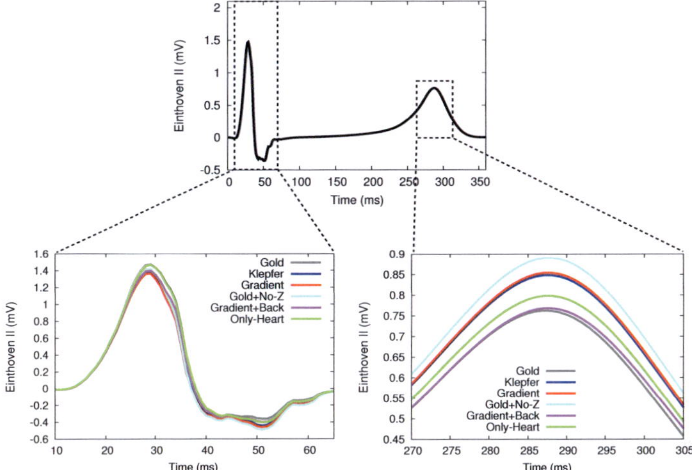

Fig. 7.9. The upper part of the figure shows the Einthoven II ECG calculated with the gold standard fiber orientation. The lower part shows a magnification of the QRS complex and the T-Wave allowing a visual evaluation of the performance of the simplified setups. In case of the QRS complex only minor differences are visible between the different setups and the gold standard. With respect to the T-Wave, most simplified setups (except Gradient+Back) overestimated the signal amplitude.

ment. However, it was unexpected to see that the best results were associated with the setup that neglected skeletal muscle fiber orientation entirely (Only-Heart).

7.3.3 Discussion

Although several studies have shown that skeletal muscle fiber orientation is important for a realistic solution of the forward problem of electrocardiography [240, 243, 15], there are up to now no established methods to extract this information from clinically acquired image data (e.g. MRI or CT). Therefore, most studies that consider skeletal muscle fiber orientation include the fiber direction based on simplified rule-based approaches.

In this study, we tried to evaluate these rule-based approaches by comparing the associated forward calculated BSPMs with the results that were based on a highly detailed gold standard fiber orientation. All rule-based approaches (except Gradient+Back) neglected the longitudinal fiber orientation components which resulted in an underestimation of the signal amplitudes in the Einthoven I lead as well as in an overestimation of the Einthoven II amplitudes. We attribute this effect to the

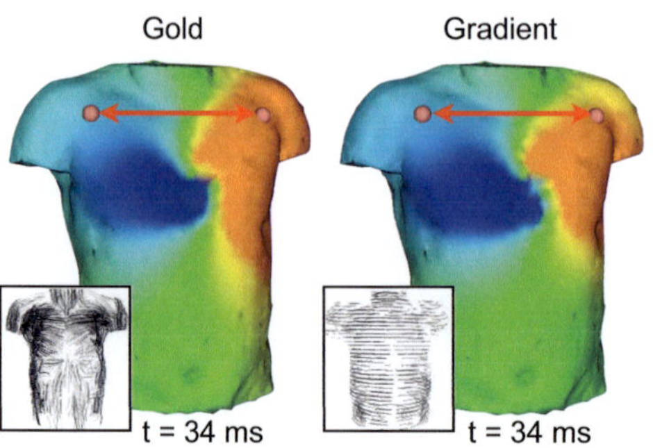

Fig. 7.10. Exemplary BSPM patterns are shown for the gold standard (Gold) and a rule-based (Gradient) fiber orientation setup during the QRS complex ($t = 34\ ms$). It can be seen, that the BSPM patterns are broadened in the direction of the fiber orientation thus influencing the signal amplitude of the different ECG leads. In this case, the electrodes are used to extract the ECG of the Einthoven I lead.

relation between the projected direction of the different Einthoven leads and the orientation of the skeletal muscle fibers. In case of the Einthoven I lead, the fibers of the rule-based setups lay parallel to the projected lead direction due to the neglect of the longitudinal components. As the larger conductivity along the muscle fibers led to a broadening of the BSPM peaks in fiber direction, the signals at the two electrodes of the Einthoven I lead became more similar, which finally reduced the amplitude of the difference signal (see Fig. 7.10). Regarding Einthoven II, the absence of a longitudinal fiber component led to a constriction of the BSPM peaks and thus to an increase in signal amplitude (see Fig. 7.11).

It is interesting to note that all rule-based approaches performed worse than the setup that completely neglected skeletal muscle fiber orientation (see Table 7.2). The reason for this was the omission of the longitudinal component. If we added longitudinally oriented back muscles (Gradient+Back), the results were at least partially enhanced. A comparison between the RMSE of the gold standard without longitudinal orientation (Gold+No-Z) and the RMSE of the rule-based approaches, however, showed that the assumption of the circumferential orientation was valid and that there was no significant inclination of the muscle fibers towards the torso's interior.

We would like to emphasize that the results of this study depended on the skeletal muscle anisotropy ratio that was used (3:1). In the literature, there are reports of much larger anisotropy ratios (up to 15.3:1, see [350, 15] and section 7.5.1). The

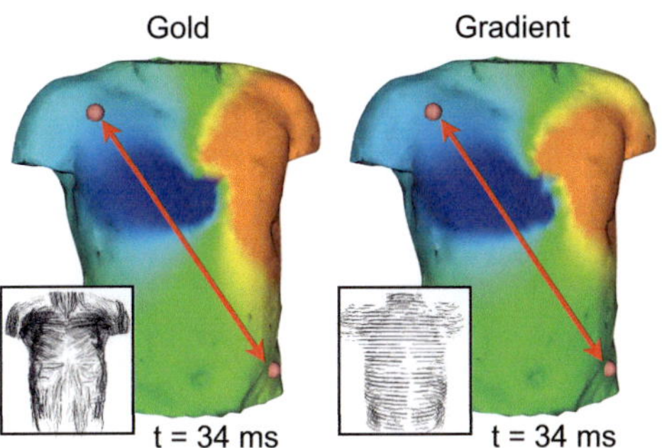

Fig. 7.11. Exemplary BSPM patterns are shown for the gold standard (Gold) and a rule-based (Gradient) fiber orientation setup during the QRS complex ($t = 34\ ms$). It can be seen, that the BSPM patterns are broadened in the direction of the fiber orientation thus influencing the signal amplitude of the different ECG leads. In this case, the electrodes are used to extract the ECG of the Einthoven II lead.

BSPM signal broadening effects in fiber direction (see Fig. 7.10 and Fig. 7.11) will increase with higher anisotropy ratios and thus the associated lead-dependent changes (see Fig. 7.8 and Fig. 7.9).

In general it can be concluded, that a simplified rule-based approach to include the skeletal muscle anisotropy as used in [243] or [381] can induce large errors in the calculation of the BSPMs. This conclusion is probably equally valid for the so-called McFee approximation to consider skeletal muscle fiber anisotropy [382]. This approximation assumes, that the anisotropic properties of the skeletal muscle layer can be modeled by increasing the thickness of the muscle layer by a factor of 3-7. Due to its fast and simple implementation, the McFee approximation has been used in a number of different studies [382, 383, 256, 384, 385, 386] most of the time even with a fixed skeletal muscle layer thickness. The analogy between the McFee approximation and the rule-based approaches which assume that all skeletal muscles fibers are oriented parallel to the torso surface is evident. Although the McFee approximation does not assume all fibers to be horizontally aligned, it does not consider potential longitudinal orientations in return (as no real anisotropy is considered and the anisotropy is modeled by simple scaling of an isotropic skeletal muscle layer, the isotropic conductivity in the plane parallel to the torso surface can be interpreted as a random fiber orientation).

Until methods are available that allow to consider the regionally heterogeneous longitudinal component of the skeletal muscle fibers (e.g. DTMRI or more sophisticated rule-based methods) it seems to be better to entirely neglect their anisotropic influence.

7.4 Results: Modeling Ventricular Deformation and the Effects on the ECG

7.4.1 Elastic 3D Image Registration

Fig. 7.12A and 7.12B show the distance of the landmarks to their target positions and the distance of all other points to their closest neighbors during the registration process from the diastolic to systolic state. With increasing iterations, the distances became smaller until the termination criterion was reached. In all cases, the termination criterion was reached after 8-12 iterations.

Fig. 7.12C and 7.12D give an overview over the landmark distances to their target positions and the distance of all other points to their closest neighbors before and after the registration procedures. The mean distances before the registration give an impression on the contraction (growing distances to the diastolic state) and relaxation (shrinking distances to the diastolic state) of the ventricles over time. As expected, ventricular deformation was a non-linear process in which the contraction phase was shorter than the relaxation phase. The standard deviation of the distances was a measure of the regional differences in deformation (i.e. some areas did undergo significantly larger displacements than others).

The rotation of the landmarks during contraction was quantified by calculating their rotation around the longitudinal axis of the left ventricle. Maximal and minimal systolic rotation was $10.7°$ and $0.6°$, respectively, while the average rotation was $4.7° \pm 2.5°$. This was in the range of reported values at the respective ventricular locations [258].

Fig. 7.13 shows the mechanical state of the DYNALAND model at 15 selected phases. The displacement vectors indicate the trajectories of the cardiac nodes. They were extracted from the deformation fields that resulted from the 29 registration procedures. Maximal systolic contraction was imaged in phase 14 as seen in Fig. 5.7 and Fig. 7.13. The Cine data was imaged using retrospective ECG gating thus beginning with the R-peak in the ECG. R-peak maximum was reached after approximately 30 ms (see Fig. 7.14). Thus, maximal ventricular contraction was imaged approximately 281 ms after the R-peak.

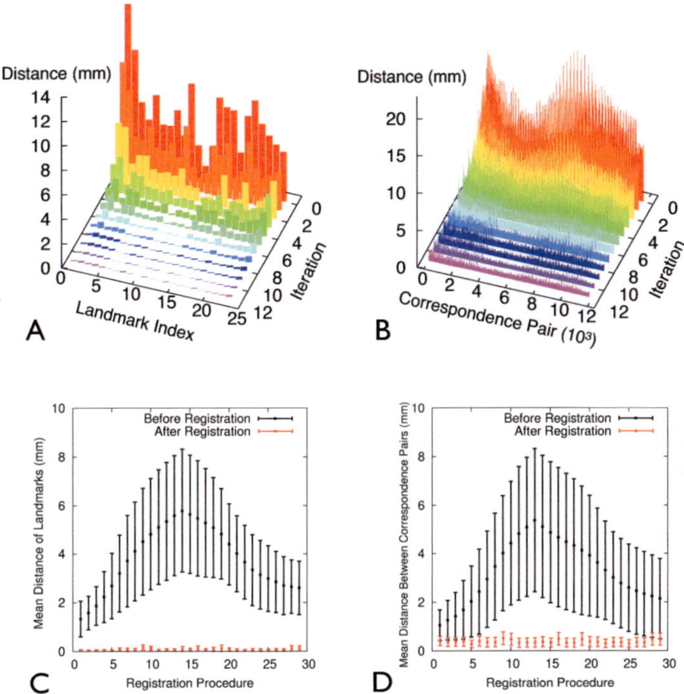

Fig. 7.12. A,B: Shown are distances during the registration from diastolic to systolic state. A: Distance of each of the 23 landmarks (DYNALAND model) to their target positions during the registration process. B: Distances of all points but the landmarks (DYNALAND model) to their closest neighbors during the registration process. After 11 iterations, the termination criterion was reached. C: Mean distance of the landmarks (± standard deviation, DYNALAND model) from their target positions before and after each of the 29 registration procedures. D: Mean distance of all points but the landmarks (± standard deviation, DYNALAND model) to their closest neighbors before and after the 29 registration procedures.

7.4.2 Electrophysiological Modeling

The ventricles were activated at the semi-automatically determined endocardial stimulation sites (see Fig. 5.13A). In order to reproduce the early activation of the basal anterior paraseptal wall that was reported by Durrer *et al.*, we relocated one of the root-points that marked the connections of the bundle branches with the Purkinje fiber network to the corresponding area. The final position of the root-points can be seen in Fig. 5.13A. Apart from this area of early activation, the

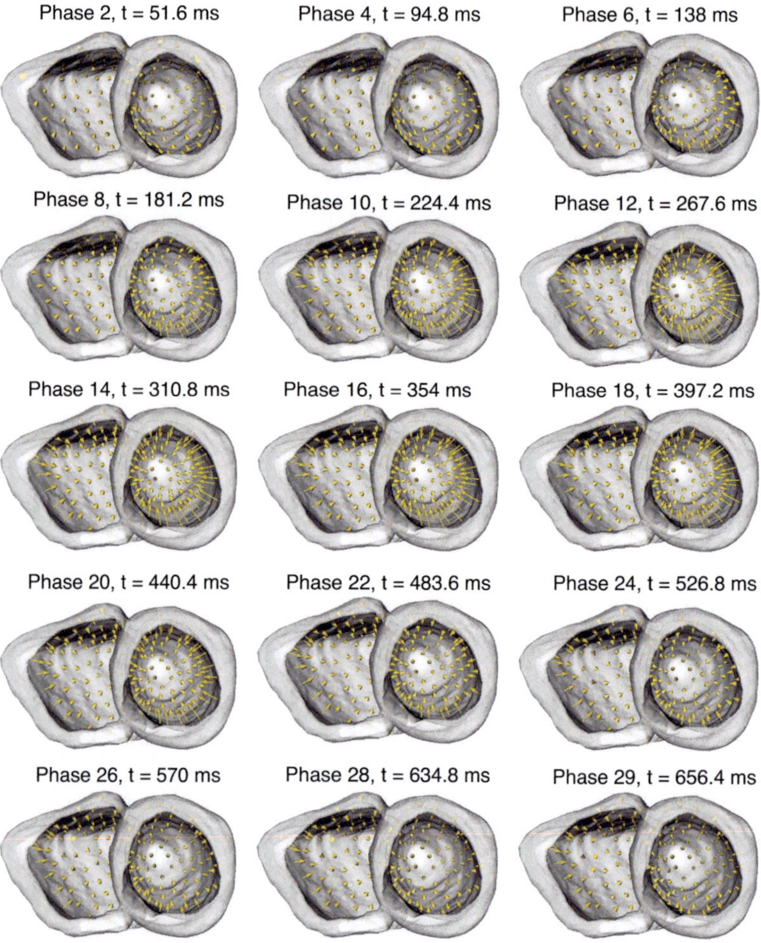

Fig. 7.13. Shown are the displacement vectors that describe the trajectories of the cardiac nodes in the DYNALAND model together with a transparent visualization of the ventricles in the diastolic state. The displacement vectors were extracted from the deformation fields that resulted from the 29 registration procedures. Phase 14 represents the state of maximal ventricular contraction.

depolarization wave spread from apex to base. Moreover, it was identical for all three electrophysiological setups (physiological QT time, SQT and LQT).

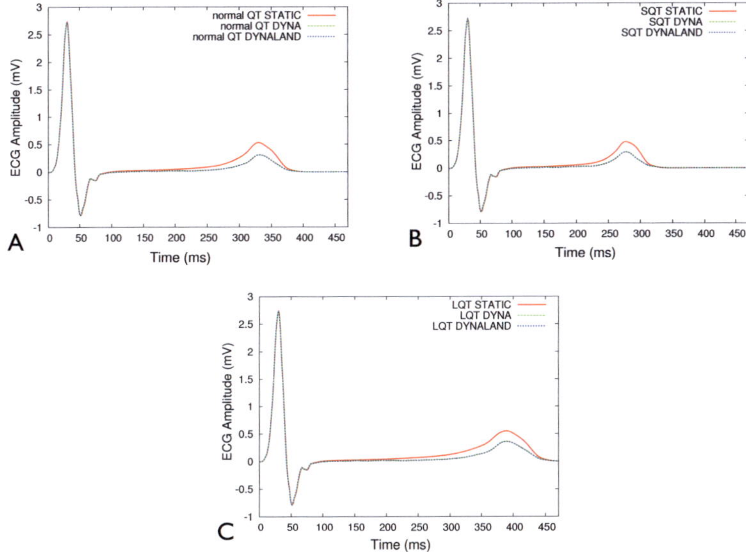

Fig. 7.14. The amplitude reductions when using the dynamic models for the forward calculations can be seen for the physiological setup with normal QT time (A) and the setups for SQT (B) and LQT (C), respectively. All plots (A-C) show the ECG in the Einthoven II lead. No significant differences were observed between the DYNA and DYNALAND model.

The sequence of repolarization was influenced by the heterogeneous ion channel distribution that was present in the model (see Fig. 5.13C). Long APD_{90} was due to small values of g_{Ks} as e.g. found in basal M-cells. The largest value of g_{Ks} was found in the apical regions of the left epicardium leading to faster repolarization in that area. The distribution of action potential durations (see Fig. 5.13D) shows no signs of the layered distribution of endocardial, M or epicardial cells. This was due to the effects of electrotonic coupling which smoothens the APD_{90} in tissue. The resulting QT times (manually extracted from Einthoven II) were 389 ms for the setup with physiological QT time and 326 ms and 461 ms for the SQT and LQT setup, respectively.

7.4.3 Impact of Ventricular Contraction on the ECG

Fig. 7.14 shows the ECGs (Einthoven II) that resulted from the forward calculations with the static and dynamic models. For all three electrophysiological setups (physiological QT time, SQT and LQT), a reduction of the T-Wave amplitude was visible when using the dynamic models during the forward calculations. However, there were no visible differences between the DYNA and the DYNALAND model. Table 7.3 summarizes the differences between the maximum amplitudes of the STATIC and the dynamic models in the Einthoven II lead. In addition to the amplitude reductions, the T-Wave maximum was also slightly postponed when using the DYNA and DYNALAND model (+2 ms for physiological QT times and +1 ms for LQT).

Fig. 7.15 shows the correlation coefficient r between the STATIC and DYNA model ($r_{StaticDyna}$) and between the STATIC and DYNALAND model ($r_{StaticDynaland}$). Correlation coefficients were calculated in each electrode and over the complete length of the ECG for the three electrophysiological configurations (physiological QT time: $0 - 400$ ms, SQT: $0 - 370$ ms, LQT: $0 - 470$ ms).

In addition to that, the absolute change δ of the T-Wave maximum was calculated by subtracting the absolute value of the T-Wave maximum of the DYNA ($\delta_{StaticDyna}$) or of the DYNALAND model ($\delta_{StaticDynaland}$) from the absolute value of the T-Wave maximum of the STATIC model. The resulting δ values for each electrode were then normalized using the electrode with the maximum δ of the respective model. The normalization factors are listed in Table 7.4. They were a measure for the magnitude of the amplitude changes for each of the three electrophysiological setups.

In general, the correlation coefficient is sensitive towards changes in signal morphology such as temporal shifts etc.. In contrast to that, the normalized absolute change is a measure of T-Wave amplitude neglecting morphological alterations. No visible changes could be observed between the DYNA and DYNALAND model for both, the correlation coefficient and normalized absolute change (see Fig. 7.15).

The correlation coefficient was clearly reduced in electrodes 7-12, 23-28, 35-40, 46-52 and 58-64 in all three electrophysiological setups. The periodicity in the

Table 7.3. T-Wave amplitude reduction of the dynamic models compared to the STATIC model (Einthoven II)

	Phys. QT	SQT	LQT
DYNA	60%	66%	67.2%
DYNALAND	59.7%	66%	66.7%

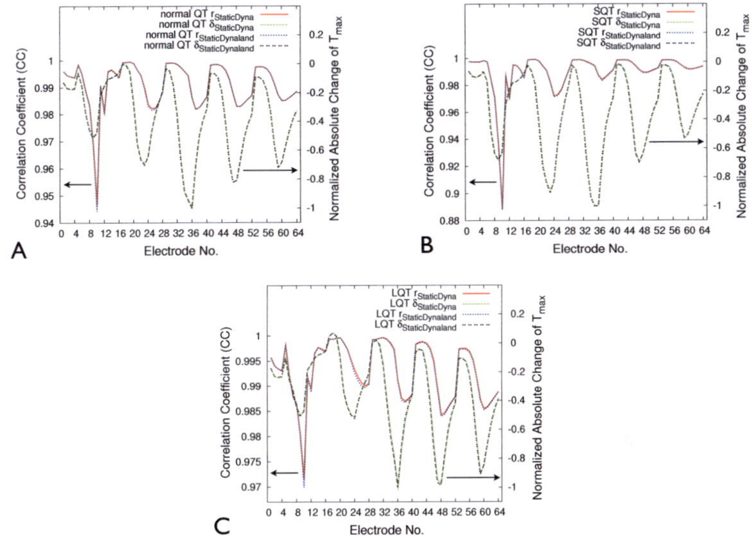

Fig. 7.15. Changes in correlation coefficient and normalized absolute change are displayed for each of the three electrophysiological setups if the dynamic models are used for the forward calculations. Both correlation coefficient and normalized absolute change are plotted over the 64 electrodes which were arranged on the torso as displayed in Fig. 5.10.

course of the correlation coefficient was clearly related to the electrode arrangement on the thorax. The electrodes were arranged in vertical stripes (see Fig. 5.10) with minima of the correlation coefficient in a region beginning at the bottom of the heart and reaching across the abdomen. Large values of the correlation coefficient were found at electrodes situated over the lungs and upper part of the torso. In case of the normalized absolute change, the region with smallest values began at the top of the heart. Values below 0 were associated with an amplitude reduction of the T-Wave maximum whereas the largest amplitude reduction could be found for normalized absolute change values of -1. While the distribution of minima and maxima of both correlation coefficient and normalized absolute change was similar in all three electropyhsiological setups, there were distinct differences when it came to the course and magnitude of the correlation coefficient and absolute change. Largest signal changes were observed for the SQT setup (both correlation

Table 7.4. Factors used to normalize the absolute change for the three electrophysiological configurations in case of the DYNA and DYNALAND model (see Fig. 7.15)

	DYNA	DYNALAND
Phys. QT	$5.12 \cdot 10^{-4}$	$5.15 \cdot 10^{-4}$
SQT	$5.53 \cdot 10^{-4}$	$5.56 \cdot 10^{-4}$
LQT	$3.54 \cdot 10^{-4}$	$3.63 \cdot 10^{-4}$

coefficient and absolute change) followed by the setup with normal QT times and the LQT setup.

7.4.4 Discussion

In this project we evaluated the effects that ventricular deformation and the movement of the associated electrical sources had on the T-Wave in the ECG. To this end, ventricular activation and repolarization was simulated using a detailed electrophysiological model that was able to consider the effects of individual ion channels [151]. Transmural and apico-basal electrophysiological heterogeneities were considered to allow for a realistic repolarization sequence. Although interventricular (between left and right ventricle) differences in ion channel density have been reported (e.g. [57]) they were not integrated in the presented model.

Ventricular activation was initiated by a special sequence of endocardially applied stimulus currents. The location and temporal sequence of these stimuli were determined based on a semi-automatic approach that was introduced in [2] and in section 4.1. Unlike the models presented by ten Tusscher *et al.* [181] and Vigmond *et al.* [180] the semi-automatic approach used here did not allow to consider the specific electrophysiological characteristics of Purkinje cell. However, it was fast and easily adaptable to different anatomical models and had a smaller number of model parameters that had to be adjusted in order to create realistic ECG waveforms. The resulting isochrone maps were in good agreement with the invasively acquired measurements from Durrer *et al.* [174].

Action potential propagation in the cardiac tissue was described with the monodomain reaction-diffusion model. Although it is a simplification of the computationally more demanding bidomain model, it has been reported to generate similar results in the absence of externally applied stimuli currents (e.g. defibrillation) [166].

In this project, we hypothesized that the magnitude of T-Wave changes will depend on synchrony/asynchrony of mechanical relaxation and electrical repolarization. Thus, it was expected that T-Waves calculated based pathologies that influenced the QT time show different changes compared to the T-Wave calculated for a physiological setup. In this case, a setup with pathologically short QT time (SQT) and a setup with pathologically long QT time (LQT) was chosen to test this hypothesis. Although the simplified models of SQT and LQT were not able to reproduce the morphological features of the pathological T-Waves, they correctly modeled the characteristic QT times for the respective pathology which was the crucial parameter for this study.

As expected, changes in the relation between mechanical contraction and relaxation and electrical repolarization led to differences in T-Wave amplitude changes. The SQT setup showed the largest T-Wave changes. In this case, the end of repolarization and the phase of maximal systolic contraction coincided. Thus, the repolarization occured during a strongly contracted state of the ventricles. For longer QT times (e.g. the setup with physiological QT time or LQT) the repolarization occurred after the systole thus resulting in smaller T-Wave amplitude changes.

Recently, the LQT syndrome has been reported to lead to longer contraction durations [387]. This would mean that the systole is prolonged and thus T-Wave changes could potentially be underestimated in our dynamic LQT models. However, we modeled a very mild form of LQT (QT_c = 461 ms). Clinically, a QT_c > 500 ms is not uncommon. Longer QT times would compensate the effects of a prolonged systole and lead to similar results as we presented here. In any case, T-Wave amplitudes were reduced in both dynamic models. This was in agreement with the results from a previous experimental [255] and in-silico study [260].

The dynamic models used to evaluate the influence of ventricular deformation on the ECG were created by elastic registration of a diastolic model to the various deformed states that were previously segmented based on Cine MRI data. We think that this approach is superior compared to approaches in which the dynamic models are created based on electromechanical modeling (see [258, 260]). This is due to the difficulties that are associated with the validation of electromechanical models. If it is not proven that the deformation is modeled correctly, conclusions regarding the impact of that deformation on the shape of the T-Wave are illegitimate. Our approach is not limited by this problem as the dynamic models used here were inherently validated based on the underlying MR images that were used for their construction.

To our knowledge, there is one other dynamic model that has been created based on MRI data and was used in a number of different studies [388, 389, 390, 259]. Yet, we think that the dynamic model presented in this work has several advantages:

- Wei *et al.* used a simplified cellular automaton to describe ventricular electro-physiology [259]. It has previously been shown that such fixed AP models have severe shortcomings in the realistic description of electrical repolarization due to the lack of electrotonic coupling [162]. In this study, we use a detailed ionic model including realistic electrophysiological heterogeneities (transmural and apico-basal) that fully accounts for the effects of electrotonic coupling.

- The MRI data used in this work had a higher spatial (factor 2) and temporal (factor 2.5) resolution. This enabled the construction of a dynamic model that captured the non-linear ventricular contraction with more accuracy.

- Wei *et al.* used a standard torso geometry to solve the forward problem of elec-trocardiography while the dynamic heart model was created based on MRI data from a volunteer. Torso and heart model were then merged prior to the forward calculations. This could potentially lead to large errors due to the strong effects of heart displacement on the body surface potentials [240]. We avoided this source of error by creating both torso and dynamic heart model from the MRI data of the same volunteer.

- Most importantly, we used a physiological QT time (389 ms) that was in agree-ment with literature data [134] and allowed to establish a realistic relation be-tween mechanical deformation and electrical repolarization. In contrast to that, we estimated a QT time slightly over 200 ms (T-Wave at 180 ms; no exact QT time given) for the study of Wei *et al.*. Even compared to pathological condi-tions like SQT this QT time is too short and thus unrealistic. In our opinion, it is not possible to draw any conclusion regarding the impact of cardiac motion on the ECG based on the model of Wei *et al.* as the physiological relation between the electrical and the mechanical processes is severely disturbed.

If we compare the dynamic model created in this study to the models that were based on electromechanical modeling ([258, 260]), it should be acknowledged that the presented approach was very labor intensive due to the manual segmenta-tion of the anatomical models and the subsequent registration procedures. In gen-eral, the spatio-temporal accuracy of our models was determined by the fidelity of the manual segmentation and the precision of the landmark positions. Manual segmentation techniques (as used in this study) are always prone to errors (e.g. individual inaccuracies and inter-individual variations). However, strategies were used to minimize segmentation errors such as the incremental segmentation procedure where phase $t + 1$ was segmented using the contours of phase t as starting basis as well as the spatial and temporal averaging of the segmented contours.

The inclusion of the manually tracked landmarks allowed a more realistic characterization of the regional heterogeneity of deformation. Yet, our model would benefit from the incorporation of additional landmarks from a larger number of slices to describe the regional heterogeneity of deformation completely. As it is unfeasible and possibly imprecise to extract a large number of landmarks manually, we envision to use the Sine Wave Modeling approach [391], which delivers displacement information from tagged MRI at sub-pixel accuracy for a large number of measurement positions without user interaction. In this case, it would no longer be necessary to segment the endocardial and epicardial surfaces manually as the registration could be performed based solely on the displacement of the automatically extractable landmarks.

One of the limitations of the dynamic model that we presented here is that it did not consider the effects of mechanoelectrical feedback as deformation was not simulated but extracted from MRI data. In the future, it would be possible to create strain maps during the registration using the biomechanical model and feed these strain maps back into the electrophysiological model to consider these effects. In a 2D model, mechanoelectrical feedback was reported to lead to a leftward shift of the T-Wave in the calculated pseudo-ECG [260]. This was attributed to the shortening of the APD due to an increase in free intracellular calcium [392]. Another limitation of the model presented in this study is that it did not consider changes in electrotonic coupling that might arise due to changes in the fiber arrangement during ventricular contraction or changes in myocardial conductivities that are associated with cell shortening [393]. Such effects could only be considered if the whole cardiac cycle would be simulated on a time dependent ventricular anatomy. However in that case, the stiffness matrix would have to be reassembled for each time instant which is computationally very expensive.

In conclusion, we have presented a realistic 3D model to evaluate the influence of ventricular contraction on the T-Wave in the ECG. In addition to its application possibilities regarding dynamic forward calculation, the created models could also be used as a reference to validate mechanical simulations in the future.

7.5 Results: Ranking the Influence of Different Tissues with Respect to their Conductivities

7.5.1 Literature Overview Regarding Conductivity Uncertainties

Fig. 7.16 lists the minimum and maximum conductivity that was found in the measurement literature. These values were used as upper and lower boundaries for the uncertainty analysis. The tabulated conductivities were sorted according to the highest ratio between upper and lower conductivity boundary. $\parallel$ and $\perp$ denote conductivities along and across muscle fibers. As explained before, we used measurement values from the colon as lower conductivity boundary for the intestine and data from the small intestine as upper conductivity boundary. To allow for a comparison to the conductivities measured by Gabriel *et al.* which were used in our standard setup, the Gabriel conductivities were listed in the table as well. For the sensitivity analysis, the conductivities from Gabriel *et al.* were varied by $\pm 25\%$. The resulting conductivities are not explicitly stated as they can be easily calculated from the tabulated values.

The intracellular transverse conductivity $\sigma_{i\perp}$ of the heart muscle and the associated anisotropy ratio were set to the values reported by Colli Franzone *et al.* [394] ($\sigma_{i\perp} = 0.031525$ S/m; anisotropy ratio: 9.516). Extracellular anisotropy ratios that are frequently reported are 2.23 [350], 2.5 [395] or 4 [396]. For the uncertainty analysis, the heart muscle conductivity at random fiber orientation was chosen for the extracellular conductivity across the fiber. The extracellular anisotropy ratio was set to 3 which was in good agreement to the previously stated ratios.

7.5.2 Sensitivity and Uncertainty Analysis

The results of the sensitivity and uncertainty analysis are shown in Table 7.5 and Table 7.6 for the atrial and ventricular input signal. Both tables are sorted according to the RMSE of the not-normalized signals. Organ ranks for a CC or RMSE$_{\mathrm{norm}}$-sorted table are given in brackets.

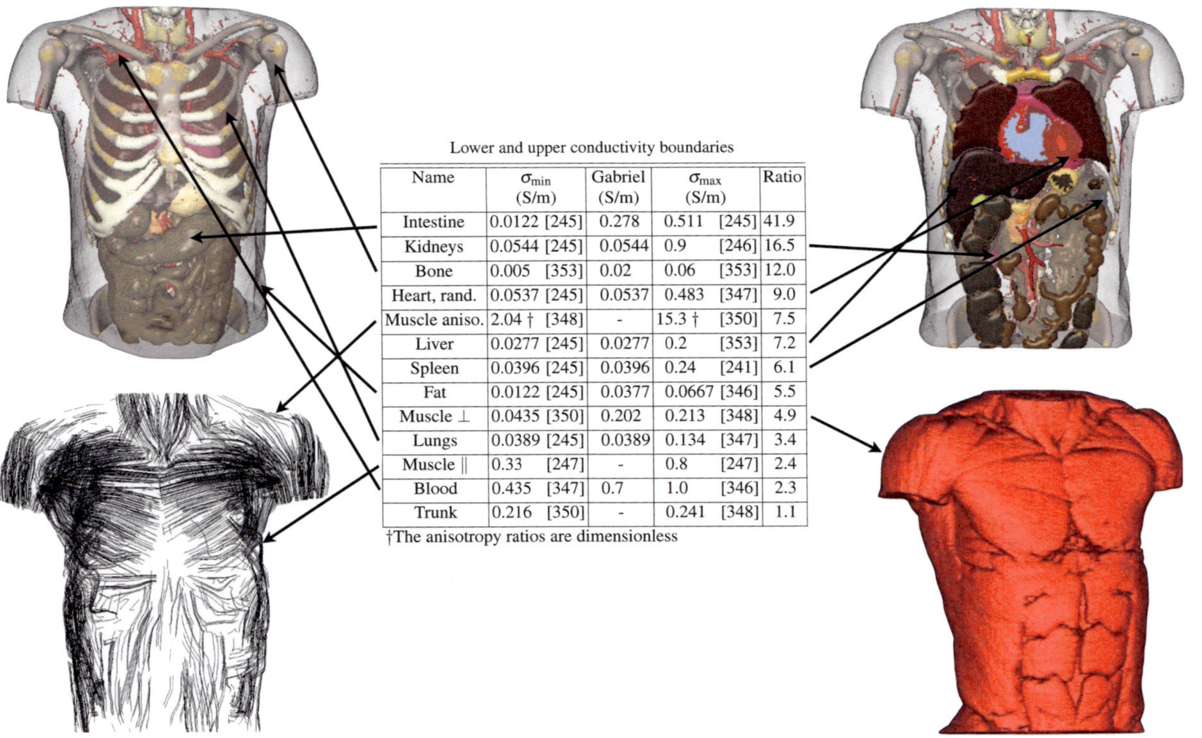

Lower and upper conductivity boundaries

Name	σ_{min} (S/m)	Gabriel (S/m)	σ_{max} (S/m)		Ratio
Intestine	0.0122 [245]	0.278	0.511	[245]	41.9
Kidneys	0.0544 [245]	0.0544	0.9	[246]	16.5
Bone	0.005 [353]	0.02	0.06	[353]	12.0
Heart, rand.	0.0537 [245]	0.0537	0.483	[347]	9.0
Muscle aniso.	2.04 † [348]	-	15.3 †	[350]	7.5
Liver	0.0277 [245]	0.0277	0.2	[353]	7.2
Spleen	0.0396 [245]	0.0396	0.24	[241]	6.1
Fat	0.0122 [245]	0.0377	0.0667	[346]	5.5
Muscle $\perp$	0.0435 [350]	0.202	0.213	[348]	4.9
Lungs	0.0389 [245]	0.0389	0.134	[347]	3.4
Muscle $\parallel$	0.33 [247]	-	0.8	[247]	2.4
Blood	0.435 [347]	0.7	1.0	[346]	2.3
Trunk	0.216 [350]	-	0.241	[348]	1.1

†The anisotropy ratios are dimensionless

Fig. 7.16. Here, a visualization of the different organs and structures is shown, that can be present in a torso model used to solve the forward problem of electrocardiography. The integrated table lists the lower and upper conductivity boundaries that can be found in the measurement literature for the respective organs. The ratios listed in the last column are a measure for the uncertainty of the measurements. To allow a comparison to the conductivities measured by Gabriel *et al.* which we used in our standard setup, the corresponding conductivities are included into the table as well.

Table 7.5. Results of the sensitivity analysis for atria and ventricles. Shown are three different rankings depending on different evaluation criteria (RMSE, $RMSE_{norm}$ and 1-CC). The numbers in brackets denote the ranks with respect to the alternate ranking criteria.

	Name	RMSE (μV)	$RMSE_{norm}$	1-CC
Atria	Muscle $\perp$	13.2	6.85E-3 (4)	2.88E-3 (6)
	Blood	10.4	1.03E-2 (2)	7.89E-3 (1)
	Muscle aniso.	9.0	6.59E-3 (6)	3.15E-3 (4)
	Lungs	6.9	1.09E-2 (1)	7.80E-3 (2)
	Fat	5.3	8.77E-3 (3)	6.14E-3 (3)
	Heart	4.6	6.19E-3 (7)	3.12E-3 (5)
	Intestine	4.1	6.82E-3 (5)	2.47E-3 (7)
	Liver	2.2	3.59E-3 (8)	9.78E-4 (8)
	Bone	1.6	2.49E-3 (9)	5.16E-4 (9)
	Cartilage	0.55	9.33E-4 (10)	6.18E-5 (10)
	Spleen	0.16	2.65E-4 (11)	5.82E-6 (11)
	Kidneys	0.12	2.01E-4 (12)	2.19E-6 (12)
Ventricles	Blood	55.1	1.70E-2 (1)	2.76E-2 (1)
	Muscle $\perp$	47.9	8.10E-3 (3)	4.20E-3 (4)
	Heart	46.1	1.07E-2 (2)	9.90E-3 (2)
	Muscle aniso.	36.7	7.80E-3 (4)	3.70E-3 (5)
	Fat	26.4	7.00E-3 (5)	4.90E-3 (3)
	Lungs	23.3	5.80E-3 (6)	3.10E-3 (6)
	Intestine	17.8	4.80E-3 (7)	1.10E-3 (7)
	Bone	9.06	2.90E-3 (8)	7.69E-4 (8)
	Liver	8.44	2.60E-3 (9)	5.90E-4 (9)
	Cartilage	4.94	1.60E-3 (10)	2.08E-4 (10)
	Spleen	0.42	1.22E-4 (11)	1.50E-6 (11)
	Kidneys	0.40	1.10E-4 (12)	8.78E-7 (12)

7.5.2.1 RMSE-based Ranking

When we compare the RMSE-sorted atrial sensitivity ranking (ASR) with the ventricular sensitivity ranking (VSR) the main differences between both rankings was the position of the heart and lungs. While the lungs were more important for atrial signals, the heart tissue ranked higher for ventricular input data. Common features of both rankings were that blood and skeletal muscle conductivity had the largest impact. Furthermore, some organs that had similar RMSE values had just swapped order in the ASR compared to the VSR (e.g.: skeletal muscle & blood, lungs & fat, bone & liver).

The main differences between the atrial uncertainty ranking (AUR; Table 7.6) and the ASR were an increase in heart and intestine importance and a decrease in blood and fat importance. Changes in the ventricular uncertainty ranking (VUR) were similar. Here, we also observed a higher rank for heart and intestine and a lower rank for blood and lung tissue compared to the VSR. When we compared the AUR with the VUR, it was again evident that the lungs were considerably more important for atrial simulations whereas the heart muscle was more important for the ventricles.

7.5.2.2 CC-based Ranking

The main differences of the CC-based ranking compared to the RMSE-based ranking were the higher importance of lung and fat tissue and the reduced significance of skeletal muscle conductivity in the ASR. In case of ventricular input signals (VSR), similar observations were made: the impact of fat tissue increased and the effects of skeletal muscle conductivity changes were reduced. All other organs in both ASR and VSR did not move at all or only by one rank.

In case of a CC-sorted AUR, skeletal muscle conductivity and anisotropy as well as blood were less important whereas heart tissue, fat and liver were more important. Regarding the VUR intestine and skeletal muscle conductivity had a lower rank yet blood, fat and bone had a higher rank. Again, all other organs showed only minor changes if AUR and VUR were compared between RMSE and CC dependent sorting.

The impact of different organ conductivities on forward calculated ECGs based on atrial or ventricular input data can also be evaluated based solely on the CC-sorted ranking. In this case, it was obvious that the lungs strongly affected atrial signals while the heart and skeletal muscle conductivity were more important for ventricular signals (ASR vs. VSR). While comparing the CC-based AUR with the corresponding VUR it was observed, that fat tissue, lungs, intestine and liver conductivity were more important for atrial signals than blood, bone, skeletal muscle conductivity and anisotropy which had larger effects on ventricular data.

Table 7.6. Results of the uncertainty analysis for atria and ventricles. Shown are three different rankings depending on different evaluation criteria (RMSE, $RMSE_{norm}$ and 1-CC). The numbers in brackets denote the ranks with respect to the alternate ranking criteria.

	Name	RMSE (μV)	$RMSE_{norm}$	1-CC
Atria	Muscle $\perp$	46.1	2.12E-2 (7)	2.43E-2 (7)
	Muscle aniso.	35.7	2.52E-2 (5)	4.49E-2 (5)
	Heart	22.0	3.28E-2 (2)	7.62E-2 (1)
	Lungs	20.9	3.25E-2 (3)	5.88E-2 (3)
	Intestine	19.6	3.32E-2 (1)	5.32E-2 (4)
	Blood	17.0	1.67E-2 (8)	2.06E-2 (8)
	Fat	16.8	2.76E-2 (4)	6.29E-2 (2)
	Liver	12.8	2.12E-2 (6)	3.40E-2 (6)
	Bone	7.0	1.12E-2 (9)	1.03E-2 (9)
	Kidneys	0.85	1.50E-3 (10)	1.24E-4 (10)
	Spleen	0.61	9.95E-4 (11)	8.16E-5 (11)
Ventricles	Heart	156.8	4.12E-2 (1)	1.71E-1 (1)
	Muscle $\perp$	156.0	2.24E-2 (5)	3.83E-2 (5)
	Muscle aniso.	145.3	2.99E-2 (2)	5.53E-2 (3)
	Intestine	91.2	2.30E-2 (4)	2.37E-2 (6)
	Blood	89.3	2.66E-2 (3)	7.06E-2 (2)
	Fat	80.9	2.23E-2 (6)	4.97E-2 (4)
	Liver	43.0	1.21E-2 (8)	1.40E-2 (8)
	Lungs	41.6	9.70E-3 (9)	9.10E-3 (9)
	Bone	40.7	1.31E-2 (7)	1.56E-2 (7)
	Kidneys	2.62	7.17E-4 (10)	4.15E-5 (10)
	Spleen	1.73	4.78E-4 (11)	2.24E-5 (11)

7.5.2.3 $RMSE_{norm}$-based Ranking

Now we compare the $RMSE_{norm}$-based ranking to the CC-based ranking as both criteria focus primarily on signal morphology rather than on mere amplitude scaling. In case of the $RMSE_{norm}$-based ASR, the skeletal muscle anisotropy and the heart tissue conductivity became less important. In contrast to that, intestine and skeletal muscle conductivity became more important. Changes in the VSR were limited to the fat tissue conductivity, which ranked lower compared to the CC-based sorting. All remaining tissues showed only minor changes in both ASR and VSR. Similar trends were seen in the AUR and VUR. In both rankings, the intestine ranked higher whereas fat ranked lower in the $RMSE_{norm}$-based ranking.

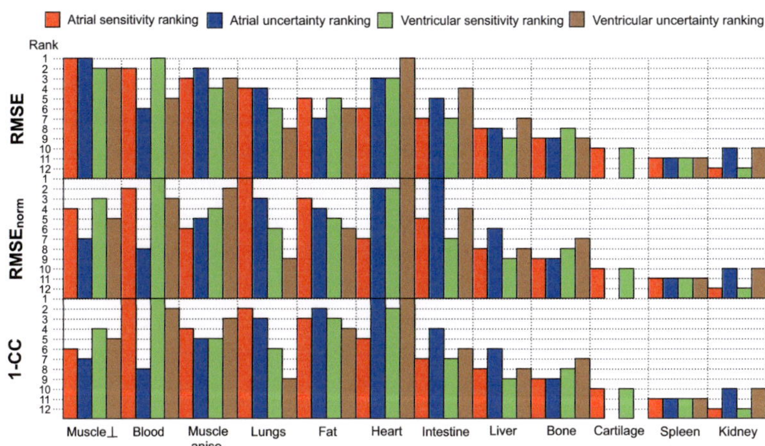

Fig. 7.17. Graphical visualization of the conductivity rankings from Table 7.5 and Table 7.6. Higher ranks in the respective categories are indicated by larger bars.

Finally, we also compared the ASR and VSR based solely on the RMSE$_{norm}$-sorted ranking: In this case lung, intestine and fat tissue conductivity had a stronger influence on atrial signals whereas skeletal muscle anisotropy and heart muscle conductivity turned out to be more important for ventricular simulations. When comparing the RMSE$_{norm}$-sorted AUR and VUR, it was observed that intestine, lungs, fat and liver were more important for the atria while skeletal muscle conductivity and anisotropy as well as blood and bone had a larger impact on ventricular BSPMs.

Fig. 7.17 summarizes the results of the conductivity rankings in a graphical visualization. Shown are both the sensitivity and uncertainty ranking for all three different comparison metrics (RMSE, RMSE$_{norm}$ and CC).

7.5.3 Possible Torso Model Simplifications

The results of the torso model simplification are shown in Table 7.7. As expected, the more organs and structures were considered, the better was the match of the associated BSPMs with the reference data. The only exception to this rule the TOP3$_{RMSE}$ setup for the ventricles based on $\overline{\sigma}_1$. In this case, the results of the simpler TOP3$_{RMSE}$ setup were closer to the reference data than the result of the more complex TOP5$_{RMSE}$ model.

Fig. 7.18 gives an impression of the differences between the simplified models and the reference setup in case of the Einthoven II lead. It can be seen, that an overly simplified torso model can lead to severe distortions of the resulting body surface signals. As a general trend, the simplified models always lead to lower R-peak amplitude. Effects on the S-peak and T-Wave were more complex. Here, the simplified models showed both smaller or larger amplitudes. However, there was a relation between changes in S-peak and T-Wave: a model that led to a larger S-peak also had a larger T-Wave in most cases and vice versa.

In addition to that, it seemed to be important which conductivity was used as replacement for the removed organs. However, no clear answer can be given regarding the best average conductivity $\overline{\sigma}$: While $\overline{\sigma}_1$ often led to better result if only a few organs were removed, a torso model filled with $\overline{\sigma}_2$ showed better results in case of completely homogeneous models.

7.5.4 Discussion

In this project, we ranked the influence of tissue conductivities on forward-calculated ECGs and evaluated the error that was associated with simplified torso models that neglected some of the lower-ranking organs. Studies with similar aims have been conducted in the past [240, 243, 397, 383]. However, this study delivers new and additional insights as it differs from previous work regarding the cardiac source distribution that was used as input for the forward calculations, the implemented anisotropy, the highly detailed torso model and the methodology that was

Table 7.7. Error of possible torso simplifications. The simplified torso models were based on the RMSE or CC-based ranking tables. The simplified models for atria and ventricles were not necessarily identical.

		Atria		Ventricles	
		$\overline{\sigma}_1$	$\overline{\sigma}_2$	$\overline{\sigma}_1$	$\overline{\sigma}_2$
RMSE	$TOP7_{RMSE}$	6.0 μV	7.1 μV	40.9 μV	46.8 μV
	$TOP5_{RMSE}$	11.2 μV	11.6 μV	55.4 μV	48.7 μV
	$TOP3_{RMSE}$	15.3 μV	16.9 μV	44.3 μV	69.8 μV
	HOM_{RMSE}	89.3 μV	42.9 μV	267.5 μV	139.4 μV
1-CC	$TOP7_{1-CC}$	7.8E-3	1.1E-2	1.0E-2	1.6E-2
	$TOP5_{1-CC}$	2.4E-2	2.9E-2	1.5E-2	2.0E-2
	$TOP3_{1-CC}$	5.0E-2	4.2E-2	5.1E-2	6.9E-2
	HOM_{1-CC}	1.3E-1	1.2E-1	1.6E-1	1.4E-1

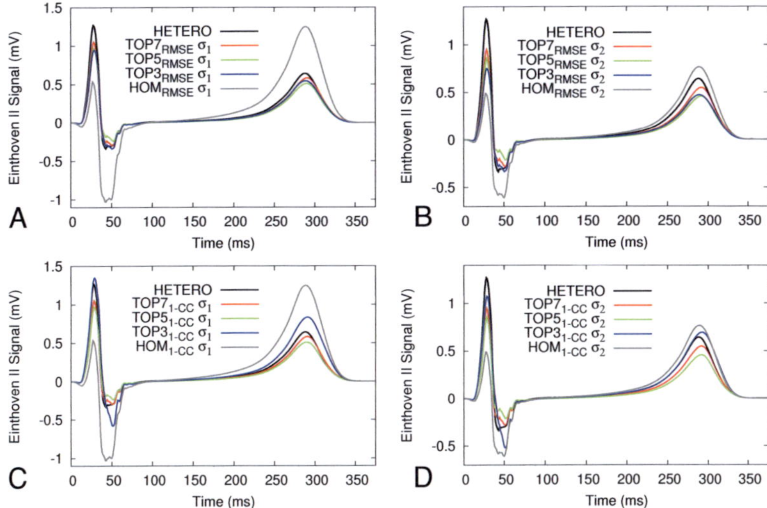

Fig. 7.18. Einthoven II lead simulated with differently simplified torso models. A,B: Simplifications were based on the RMSE-sorted ranking. C,D: Results of the simplified models if the removed organs were chosen based on the CC-sorted ranking. The HETERO setup was used as reference and denoted the fully heterogeneous model that contained all organs and structures with their respective conductivities. The quality of the simulation results with the simplified models also depended on the average torso conductivity that was used to replace the removed organs.

used to evaluate the impact of the different conductivities. The most important differences are:

- To the best of our knowledge, this is the first study that investigated the role of different torso conductivities based on *simulated* input signals of atrial and ventricular activation and repolarization. This means that our results are especially relevant regarding the future use of cardiac simulations in clinical applications. In contrast to that, previous studies were based on dipole [240, 397, 256] or double layer sources [384, 385], sources that were calculated using an inverse procedure [398] or epicardial recordings that were acquired from intact dog [383], a Langendorff-perfused dog heart [386] or during open heart surgery on a human [243].

- The skeletal muscle fiber orientation used in this study was very realistic as it was derived from the thin-section photos of the Visible Man dataset [253]. However, all other studies except for the work of Bradley *et al.* [240] used rule-based approaches [243], an approximation proposed by McFee *et al.* [383, 256, 384, 385, 386, 382] or neglected anisotropic influences at all [397, 398]. However, in case of the McFee approximation it has been suspected [240] and for the rule-based approach [243] we have previously shown [12] (and section 7.3) that such methods are not able to produce realistic anisotropy information. This is particularly problematic, as the skeletal muscle anisotropy was found among the 4 most important organs in any of the four presented rankings (ASR, AUR, VSR, VUR).

- Due to computational limitations, there are significant differences with respect to the details of the torso model that was used for the conductivitiy evaluations. While the eccentric spheres models that were used in [384] and [385] have been criticized by Bradley *et al.* [240] and van Oosterom *et al.* [398] and the use of "tailored" geometries was favored, several of the older studies used only coarse torso models with a few hundred or thousands of nodes [397, 256, 398, 386]. Regarding the level of detail that is considered in the different torso models, we would like to stress that even the most recent studies [240, 243, 386] evaluated only a subset of the organs and structures that are analyzed in this study. A low number of organs can lead to potential errors in a conductivity evaluation study, as it has been reported [243] that a simple additive relationship between the different structures is unlikely. On the one hand this means that the removal of organs need not in any case reduce the quality of the results (as observed in this study for the $TOP3_{RMSE}$ vs. $TOP5_{RMSE}$ setup in Table 7.7) as the effects of some structures might cancel each other out. On the other hand, it can be difficult to judge the importance of organs in a simple torso model that only considers a small number of inhomogeneities as the results might have been different if additional structures would have been considered. Thus the results from investigations that evaluated torso models of different detail are not necessarily directly comparable.

- When we look at previous simulation studies (and also on studies that evaluated the impact of tissue conductivities in the torso model), it is obvious that there are severe differences in the utilized conductivities (and in the ratios between the conductivities of the different organs). Yet, none of these studies tried to estimate the effects that these conductivity uncertainties might have had on the result.

In this study, we therefore differentiated between a sensitivity and an uncertainty analysis. For the sensitivity analysis we used fixed percental changes of the conductivities published by Gabriel *et al.*. In order to evaluate the influence of measurement uncertainties on the body surface potentials, we first collected literature values for each organ and then conducted forward calculations for each organ's minimum and maximum conductivity.

We used three different metrics (RMSE, $RMSE_{norm}$, CC) to rank the sensitivity and the influence of conductivity uncertainties for each organ. In case of the sensitivity analysis and the RMSE as evaluation criterion, it turned out, that both atrial and ventricular signals were most sensitive to changes in skeletal muscle, blood, heart, lungs, and fat conductivity as well as skeletal muscle anisotropy. When focusing on surface ECG morphology (evaluation criterion: CC), the skeletal muscle conductivity had a lower impact but fat on the other hand was more important. It is interesting to note that for all three evaluation criteria the lungs were more important for atrial simulations whereas the heart had a stronger impact on the ventricles. The reason for this is seen in the location of the atria which are completely surrounded by lung tissue. The stronger influence of heart tissue conductivity on the ventricular signals is probably due to the much thicker chamber walls of the ventricles compared to the atria.

As previously explained, it is difficult to directly compare the results of different studies that evaluated the role of conductivity variations for the forward problem. Reasons for these difficulties are methodological differences, differences in torso models and chosen conductivities. While considering these limitations, we found the following similarities and differences:

- The lungs were previously found to be both important [240, 243, 397, 384], and unimportant [398, 383].
- Skeletal muscle conductivity had a large [385], moderate [240, 384], or small [243] effect on torso potential. In contrast to that, skeletal muscle anisotropy was always considered to be important [243, 383].
- Heart tissue conductivity had a strong impact in two studies [240, 398], which is in agreement with our results.
- Our results confirmed the large effects of changes in blood conductivity as reported by Rudy *et al.* [384] and van Oosterom *et al.* [398]. Yet this contradicts earlier reports from Rudy *et al.* [385] were blood was found to be less important.

- Our study showed moderate effects of fat conductivity on signal amplitude (yet stronger effects on signal morphology). These findings are in contradiction with reports from Bradley *et al.* [240] and Klepfer *et al.* [243] where large effects were seen and findings from Rudy *et al.* [384] where only small effects were found.

- The effects of spine and sternum were always found to be small (if investigated at all) [243, 397, 383]. Klepfer *et al.* additionally evaluated the role of bone, which was also found to be of low impact. In this study, we used a torso model that included all major bony and cartilaginous structures but likewise found no significant effects.

No other study evaluated the effects of intestine, liver, spleen and kidneys. Among these organs, our results show, that the intestine was the most important structure (followed by the liver) which was probably due to its larger size: intestine: 6.3% vs. liver: 4.1% vs. spleen 0.5% and kidneys 0.8% of the total torso volume). However, these organs marked the lower end of most rankings.

If the effects of conductivity uncertainties were additionally considered, the importance of blood and fat decreased whereas the influence of heart tissue and intestine increased. These changes were directly related to the degree of uncertainty (see Table 7.6): The larger the ratio between upper and lower conductivity boundary, the higher the rank of the respective organ in the uncertainty analysis. Especially for the rarely considered organs and structures like the kidneys, bone, liver, and spleen but also for some very important tissues like the heart or skeletal muscle fiber anisotropy, there are significant uncertainties which should be addressed by the measurement community.

It is difficult to give one single recommendation regarding the level of detail that should be incorporated into a torso model that is used to solve the forward problem of electrocardiography. In general, the more organs and structures that were considered in the model, the better was the match with the completely heterogeneous reference model that served as gold standard (the only exception was the TOP3$_{RMSE}$ model for the ventricular signals). As expected, simplification results also depended on the conductivity of the removed organ. If it was closer to the average torso conductivity that was used to replace the organ, the removal had smaller effects.

For most applications however, it should be sufficient to include the 5 most important inhomogeneities in addition to the heart. According to our results, these organs and structures were skeletal muscle conductivity as well as anisotropy, blood, lungs, and fat. This list was identical for both atrial and ventricular input signals and agreed with the suggestions from Bradley *et al.* [240], although the ranking of the organ importance was different and they were not able to evaluate the effects of the intracavitary blood pool. It is also interesting to note that the solution quality of the simplified models depended on the averaged torso conductivity $\overline{\sigma}$, that was used for organ replacement. Especially for clinical applications which would benefit from the time saving that is associated with simplified models it might be possible to derive an optimized $\overline{\sigma}$ that minimizes the differences between simplified and completely inhomogeneous torso models.

The differences between the simplified torso models and the inhomogeneous reference were larger in our work compared to previous studies [240, 243]. We think that this is related to the elevated heart conductivity that both studies used to model the effects of the ventricular blood indirectly [240]. According to Bradley *et al.*, changes in BSPM pattern and amplitude are greater if the heart conductivity is lower [240].

In contrast to the results of our study, Ramanathan and Rudy [386] claim that it is not necessary to use an inhomogeneous torso model as the BSPM differences resulting from the use of a homogeneous model are small and fall in the range of normal interindividual variations. Yet, their conclusions are weakened by their simple, low resolution torso model (containing only 6684 triangular elements that described the shape of the lungs, bone and skeletal muscle (McFee approximation)), which was used in conjunction with the boundary element method to calculate transfer matrices that linked the epicardial potentials to the body surface potentials. In addition to that, none of their presented 12-lead ECGs looked physiological (notched QRS complexes, ST elevation or depression, discordant T-Waves (if any at all)). Furthermore, the only difference in the female model that was used to evaluate possible gender-specific variations was the additional incorporation of fat tissue to model the breasts. All remaining contours looked identical to the male model although the description in the method section suggested the creation of a completely unique model based on different source data. In conclusion, it should be noted that our results shown in Fig. 7.18 and Table 7.7 clearly dispute their recommendation for a homogeneous torso model.

Finally, we evaluated the effects of blood in the major vessels (superior and inferior vena cava, pulmonary arteries and veins, aorta) and the fat layer around the heart individually. In the Visible Man dataset that was used for this study, these structures were a subset of the blood and fat, respectively. However, in patient-specific models they are rarely included due to the time-consuming segmentation procedure. To evaluate their effects, we separated them from the remaining blood and fat by assigning unique tissue classes. Then, a similar sensitivity analysis was conducted as described in section 5.5.3. In case of an RMSE-sorted ASR, the blood in the main vessels had large effects (rank 4) and should be included in all torso models for atrial simulations. In contrast to that, the fat around the heart ranked only on position 8 and might be negligible. For the RMSE-sorted VSR, both fat around the heart (position 7) and blood in the main vessels (position 12) had small effects.

When examining the results of studies that evaluated the sensitivity of the BSPM regarding changes in tissue conductivities, it has to be considered that all results depend on the default conductivities and the torso model that is used. Identical default conductivities in conjunction with a different torso model or identical torso models combined with different default conductivities will produce different sensitivity rankings. Although we tried to quantify the effects that the present conductivity uncertainties have on electrocardiographic simulations we cannot solve this problem by determining the correct conductivities based on computer simulations. It will be the task of the measurement community to eliminate these uncertainties and further narrow down the possible choices.

It should be noted, that different organs have slightly varying influences on different segments in the ECG (P, QRS, T) [240, 243]. This means that different atrial or ventricular excitation sequences might produce slightly different RMSE or CC values (but not necessarily different ranking positions). For evaluation purposes, we performed atrial simulations with different conduction velocities in the Crista Terminalis and Bachmann's Bundle (both set to 1400 mm/s). With this modified anisotropy setup, the ranking positions were almost unchanged, which means that the conclusions drawn in this work should be applicable to other physiological excitation sequences as well.

Before transferring the results of this study to different torso models, effects of varying torso compositions should be kept in mind. On the one hand, there are well-known gender-specific differences with respect to the volume of skeletal muscle vs. fat tissue. On the other hand, the Visible Man dataset has a very high fraction of body fat (38% within the torso) thus results will be different for more slender models. Furthermore, the Visible Man dataset contains fine structures with a high level of detail that are currently not available in patient-specific models based on CT or MRI data. This could mean that such models can have different conductivity sensitivities.

It would be interesting to repeat this study with a larger pool of different torso models (height, weight, gender, body composition). Currently this is not possible due to the lack of methods that allow the creation of skeletal muscle fiber orientation in patient-specific datasets. If such methods become available, a similar study on a representative group of torso models would allow a concluding evaluation regarding the impact of different organ conductivities on body surface potentials

7.6 Results: BSPM Prediction for Varying Conductivities Based on PCA

7.6.1 Eigenvalue Ratios and Eigenvector Angles

Table 7.8 lists the ratios of the first and second eigenvalues λ_1/λ_2 for the four different tissues. As the minimal ratio was 31 it was obvious that for all organs, the information content of the first principal component (PC) was much larger than that of the second PC. This property was used in section 5.6.3.3 when transforming equation (5.8) to (5.9).

In addition to that, the pairwise angles between the first eigenvectors of all four tissues were in the range of 45° to 124° for the atria and 57° and 143° for the ventricles. Thus, no two tissues had parallel eigenvectors, which would imply a similar effect of the tissues during conductivity variations.

7.6.2 Signal Reconstruction

Fig. 7.19 shows the shifted PCA scores of the four different tissues for atrial and ventricular data. All score curves did monotonically depend on the conductivity

Table 7.8. Ratio between first and second eigenvalue

	Blood	Muscle	Lung	Fat
Atria	232	181	299	77
Ventricles	652	106	489	31

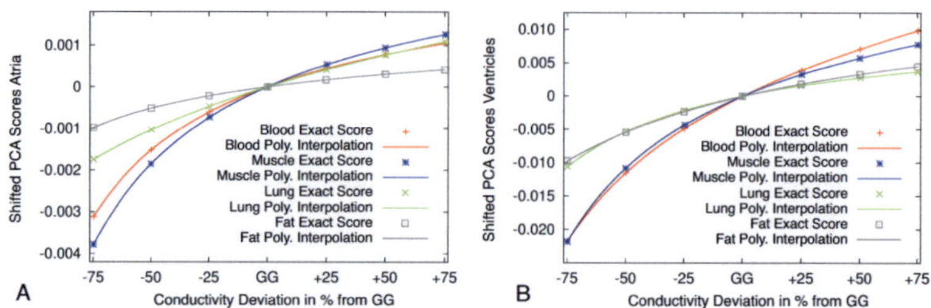

Fig. 7.19. Shifted PCA scores of the first eigenvectors for each of the four tissues in case of atrial (A) or ventricular (B) data. The data points denote the scores that directly resulted from the PCA decomposition of the forward calculated input data. The solid lines are a result of the polynomial interpolation technique. The slope of the score curve reflected the sensitivity of the forward problem to variations in the respective conductivity.

values. As all eigenvectors were normalized, the slope of the score curves was a measure for the sensitivity of the signal towards the respective tissue. Steeper slopes indicated a higher sensitivity of the reconstructed signal for a given conductivity variation.

The reconstruction errors (RMSE) associated with the prediction of conductivity changes in a single tissue are listed in Table 7.9. In case the reconstruction was based on the *exact* scores within ±75% of the GG conductivity and did only consider the first eigenvector, the maximum RMSE was in the range of $2.6\,\mu$V for the atria and $12.6\,\mu$V for the ventricles (see upper part of Table 7.9). This means, that the RMSE was at least two orders of magnitude smaller than the BSPM signal amplitudes which are in the millivolt range.

The lower part of Table 7.9 lists the errors that resulted from the reconstruction of single tissue conductivity changes based on the *interpolated* scores. In this case, the reconstructible conductivity range was limited to ±50% of the GG conductivity due to the polynomial interpolation technique. The maximum errors were in the same range ($1.8\,\mu$V for the atria and $8.5\,\mu$V for the ventricles) as the maximum errors for the exact scores (for variations of ±50%).

Table 7.10 summarizes the result for combined conductivity variations in all four organs. In case of atrial reconstruction and changes of ±25%, the average RMSE was $1.6\pm0.4\,\mu$V. As expected, the average RMSE was slightly larger for larger conductivity variations of ±50% (average atrial RMSE: $6.3\pm2.9\,\mu$V). For the ven-

Table 7.9. RMSE (μV) when using *exact* and *interpolated* scores

		Blood	Muscle	Lung	Fat
Atria Exact Scores	$\sigma_{-75\%}$	1.910	2.612	1.389	1.200
	$\sigma_{-50\%}$	0.387	0.457	0.259	0.139
	$\sigma_{-25\%}$	0.422	0.097	0.588	0.207
	σ_{GG}	0.002	0.001	0.001	0.000
	$\sigma_{+25\%}$	0.597	0.832	0.307	0.328
	$\sigma_{+50\%}$	1.248	1.695	0.751	0.715
	$\sigma_{+75\%}$	1.903	2.557	1.274	1.120
Ventricles Exact Scores	$\sigma_{-75\%}$	5.782	12.597	2.868	12.120
	$\sigma_{-50\%}$	1.009	1.335	0.343	0.767
	$\sigma_{-25\%}$	1.071	2.492	0.516	1.945
	σ_{GG}	0.008	0.002	0.002	0.003
	$\sigma_{+25\%}$	1.586	3.884	0.866	3.166
	$\sigma_{+50\%}$	3.370	8.100	1.895	6.875
	$\sigma_{+75\%}$	5.215	12.434	2.975	10.713
Atria Interp. Scores	$\sigma_{-50\%}$	0.566	0.709	0.209	0.216
	$\sigma_{-25\%}$	0.441	0.609	0.114	0.219
	$\sigma_{+25\%}$	0.598	0.831	0.296	0.324
	$\sigma_{+50\%}$	1.316	1.779	0.783	0.748
Ventricles Interp. Scores	$\sigma_{-50\%}$	1.465	2.420	0.560	1.671
	$\sigma_{-25\%}$	1.151	2.591	0.547	2.081
	$\sigma_{+25\%}$	1.585	3.862	0.864	3.113
	$\sigma_{+50\%}$	3.578	8.475	1.994	7.209

tricles, the average RMSE was $5.8\pm1.5\,\mu$V for 25% variations and $19.0\pm5.6\,\mu$V for 50% variations.

An example of a reconstructed signal that resulted from conductivity variations in all four organs can be seen in Fig. 7.20.

Table 7.10. RMSE (μV) for combined conductivity variations z in all four tissues (z={25%,50%}). Errors were larger for larger variations and in case of ventricular compared to atrial signals.

	Atria		Ventricles	
Relative Variation z	25%	50%	25%	50%
$\sigma_{B-z}\,\sigma_{M-z}\,\sigma_{L-z}\,\sigma_{F-z}$	2.1	11.3	7.2	22.6
$\sigma_{B-z}\,\sigma_{M-z}\,\sigma_{L-z}\,\sigma_{F+z}$	1.9	10.0	4.1	18.8
$\sigma_{B-z}\,\sigma_{M-z}\,\sigma_{L+z}\,\sigma_{F-z}$	2.7	13.6	8.4	27.7
$\sigma_{B-z}\,\sigma_{M-z}\,\sigma_{L+z}\,\sigma_{F+z}$	1.0	5.7	4.5	21.0
$\sigma_{B-z}\,\sigma_{M+z}\,\sigma_{L-z}\,\sigma_{F-z}$	1.7	6.7	5.0	18.1
$\sigma_{B-z}\,\sigma_{M+z}\,\sigma_{L-z}\,\sigma_{F+z}$	1.7	5.8	8.1	26.2
$\sigma_{B-z}\,\sigma_{M+z}\,\sigma_{L+z}\,\sigma_{F-z}$	1.3	5.4	4.3	15.0
$\sigma_{B-z}\,\sigma_{M+z}\,\sigma_{L+z}\,\sigma_{F+z}$	1.7	5.2	7.0	18.4
$\sigma_{B+z}\,\sigma_{M-z}\,\sigma_{L-z}\,\sigma_{F-z}$	1.2	4.3	6.0	16.6
$\sigma_{B+z}\,\sigma_{M-z}\,\sigma_{L-z}\,\sigma_{F+z}$	1.7	6.0	5.8	20.9
$\sigma_{B+z}\,\sigma_{M-z}\,\sigma_{L+z}\,\sigma_{F-z}$	1.4	5.0	7.9	26.5
$\sigma_{B+z}\,\sigma_{M-z}\,\sigma_{L+z}\,\sigma_{F+z}$	1.5	5.9	5.5	19.1
$\sigma_{B+z}\,\sigma_{M+z}\,\sigma_{L-z}\,\sigma_{F-z}$	1.3	3.7	6.0	20.8
$\sigma_{B+z}\,\sigma_{M+z}\,\sigma_{L-z}\,\sigma_{F+z}$	1.6	5.2	4.2	7.6
$\sigma_{B+z}\,\sigma_{M+z}\,\sigma_{L+z}\,\sigma_{F-z}$	1.2	3.6	4.8	15.9
$\sigma_{B+z}\,\sigma_{M+z}\,\sigma_{L+z}\,\sigma_{F+z}$	1.2	2.9	4.8	9.3
Average	1.6	6.3	5.8	19.0

7.6.2.1 Confidence Intervals

Fig. 7.21 shows the resulting confidence intervals for relative conductivity variations of $\pm10\%$, $\pm30\%$, and $\pm50\%$ in all four organs. For a larger relative uncertainty $\delta\sigma_{\mathrm{rel}}$, the gap between the GG signal and possible upper and lower signal boundaries will become wider. The relative uncertainty should be chosen conservatively such that the real signal will definitely lie within the calculated confidence intervals.

7.6.2.2 Conductivity Optimization

Table 7.11 shows an example of the conductivity optimization results. In this case, the reference for the conductivity optimization was a signal in which the conductivities of all four tissues were increased by 25%. For blood, the optimization delivered a conductivity of 0.881 S/m (compared to 0.875 S/m), which corresponded

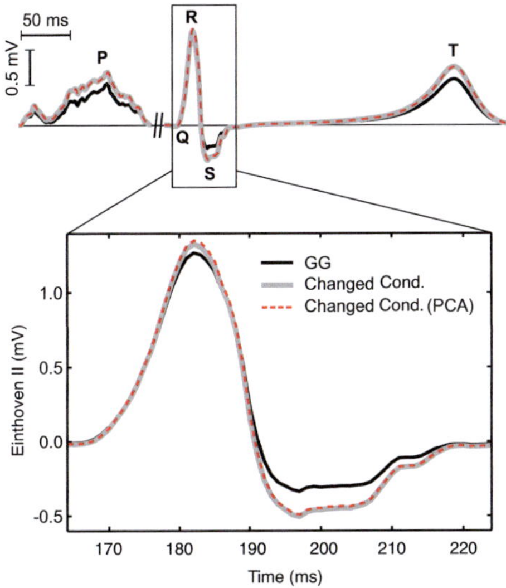

Fig. 7.20. Example of potential ECG changes due to conductivity variations. The signal in which all four conductivities were varied (here: blood and muscle decreased by 25%, lungs and fat increased by 25%) shows significant deviations compared to the default signal (GG) that was calculated with the conductivities from Gabriel *et al.*. The bottom part of the figure is a magnification of the QRS complex shown above. The PCA method presented in this study could be used to predict the effects of these combined conductivity variations in all ECG segments (P-wave, QRS complex, T-Wave). In this case, the P-wave amplitude was multiplied by two for better visibility.

to a percental increase of 25.9% (compared to 25%). So the resulting optimization error for the blood conductivity was 0.9 percentage points.

Over all 16 conductivity setups which we used to evaluate the optimization method, the averaged error (over all four tissues) was 3.0 ± 2.2 percentage points for atrial data and 4.7 ± 4.2 percentage points for ventricular data.

7.6.3 Discussion

In this study, we developed a PCA-based technique that allowed to predict changes in BSPMs that were associated with conductivity variations in a single or multiple tissues. The effects of these conductivity changes could be predicted over a wide range of conductivities from few sample simulations which were used as input for the PCA approach. The presented method was validated for both atrial and ventricular signals.

A key assumption of the method was, that changes in BSPMs due to conductivity variations in a single organ can be described with the first eigenvector of a PCA decomposition. This assumption was verified by calculating the eigenvalue ratios between the first and second eigenvalues for all four organs. As the ratio was always larger than 30, it was permissible to assume that the major part of the signal variation was described by the first eigenvector. The assumption was also indirectly confirmed by the relatively small errors that were observed for signal reconstructions (single tissue conductivity changes) based on the *exact* scores. The associated RMSE was always under $12.6\,\mu V$, thus no subsequent eigenvectors needed to be included in the reconstruction. In general, ventricular RMSEs were larger than the errors for atrial signals which we attribute to the larger ventricular signal amplitudes.

Using the proposed PCA method, only seven sample simulations (with different conductivities for each tissue) had to be performed to enable the reconstruction of BSPM signals for arbitrary conductivities within the simulated range. This was possible, as signals for values between the simulated input conductivities could be predicted by interpolating the PCA score curve. This technique was verified using a leave-one-out validation for the $\pm25\%$ and $\pm50\%$ simulations. Here, the RMSEs of the reconstructions with the *interpolated* scores were not significantly larger than the RMSEs that were associated with the *exact* scores for similar conductivity variations. It should be noted, that it was not possible to perform a leave-one-out validation for the $\pm75\%$ simulations as the polynomial interpolation was only defined between the minimum and maximum conductivity boundaries that were used as input. If a simulated value at the boundary (e.g. at -75%) would be removed from the PCA input data, the interpolated score curve would only be defined in the -50% to +75% region.

Due to the problems that are associated with the exact measurement of in-vivo conductivities (see section 3.2.3.2) and the possible inter-individual variations, there is usually more than one conductivity that is not exactly known. Therefore, we expanded our PCA-based prediction technique and demonstrated that the effects of combined conductivity variations can also be considered with the presented method. This was possible by aligning the coordinate systems of the different PCAs based on the GG signal which served as common origin. The reconstruction of the effects of combined conductivity changes was based on the assumption that the changes can be expressed by superposing the effects of the respective

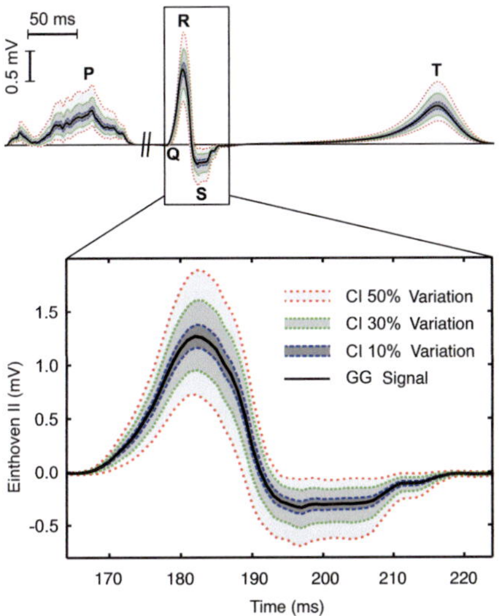

Fig. 7.21. Visualization of the signal calculated with the default conductivities according to Gabriel *et al.* (GG signal) and different confidence intervals (CI). Shown are the upper and lower signal boundaries in the Einthoven II lead that can result for different conductivity uncertainties. Simultaneous uncertainties of $\pm 10\%$, $\pm 30\%$, and $\pm 50\%$ in all four tissues were considered. Again, the P-Wave amplitude was multiplied by two to enhance signal visibility.

single conductivity variations. Although there are reports indicating that the effects of conductivity variations in two different organs are not necessarily additive [243, 399], our results (e.g. Fig. 7.20) showed that the PCA method was able to predict the signal changes in the investigated conductivity boundaries. Other indicators for the applicability of the presented method were the low average RMSE values for the combined conductivity variations of $\pm 25\%$ (atria: $1.6\,\mu V$ / ventricles: $5.8\,\mu V$). For larger variations of $\pm 50\%$, the average RMSE only rose to $6.3\,\mu V$ for the atria and $19.0\,\mu V$ for the ventricles. It should be noted that in a real clinical setting, the amount of conductivity variation is likely to be different for each tissue. However, this does not limit the feasibility of the presented approach as we only chose equal conductivity variations for the sake of simplicity.

On the other hand, there are of course signal changes due to combined conductivity variations that cannot be considered with the presented PCA method. Among these changes are non-linear cancellation effects or non-linear signal amplification. In this context, we want to emphasize that the linearization used in this study is only an approximation. In general, linear approximations are feasible for many problems. However, the range in which these approximations deliver acceptable results must be critically evaluated. Within the conductivity boundaries evaluated in this study, the RMSE values showed that the PCA method was able to provide a good estimation of the signal changes that should be expected. In contrast to the time-consuming forward calculations (duration over one hour), the PCA decomposition and reconstruction took approximately 1 s (including program startup and data loading) on a standard desktop computer using one CPU.

In addition to the reconstruction of BSPM signals for different combinations of conductivities, the PCA method can also be used to estimate confidence intervals for arbitrary ECG leads without having to perform additional forward calculations. To this end, we used the reconstruction equation to calculate the minimum and maximum signal that could occur within given conductivity boundaries (for each time step). Due to the monotonic relation between conductivities and PCA scores, only the signals at the boundaries of the conductivity ranges needed to be evaluated. It should be noted, that the resulting confidence interval curves (e.g. Fig. 7.21) were normally not based on a single conductivity setup. They can rather be interpreted as a synthetic signal as different conductivity combinations led to the minimal and maximal signals at different time instants. This ensured, that the confidence interval curves defined a range in which the real ECG signal will in all cases lie (despite potential uncertainties in the tissue conductivities).

The presented technique to calculate confidence intervals could be expanded to allow to predict the *probability* for the reconstructed signal to lie within certain boundaries, similar to [399]. In this case, the PCA method would be used to predict the signal at finite conductivity steps between the upper and lower conductivity boundaries. The results could then be statistically evaluated e.g. by calculating the mean signal and standard deviations. Although such a statistical evaluation would not be as detailed as the approach proposed in [399], it would have the advantage that the signals could be calculated *a posteriori* (without having to change the setup for the forward calculation). In addition to that, it would be possible to apply the PCA approach to a 3D torso model and at all time steps, which is currently not possible with the highly sophisticated method proposed by Geneser *et al.* [399].

As it was possible to fit a polynomial function to describe the dependency of the PCA scores on the different tissue conductivities, it was also possible to use this interpolated function in the opposite direction. In this case, it was not used to de-

Table 7.11. Optimization example for atrial data

	Blood	Muscle	Lung	Fat
GG Conductivity (S/m)	0.7	0.202	0.0389	0.0377
Percental Increase (Reference)	+25	+25	+25	+25
Reference Conductivity (S/m)	0.875	0.253	0.0486	0.0471
Percental Increase (Estimated)	+25.9	+23.8	+19.5	+20.0
Estimated Conductivity (S/m)	0.881	0.250	0.0465	0.0453
Abs. Error in Percentage Points	0.9	1.2	5.5	5.0

rive PCA scores for conductivities in-between the forward calculated input data but rather to determine the conductivity for a given PCA score. Nelder-Mead's simplex optimization scheme [364] was used to probe hundreds of different conductivity combinations and the resulting BSPMs were compared with the reference signal. The PCA scores of the best matching signal were then used to determine the corresponding conductivities. This kind of brute-force optimization would normally not be possible with the standard forward calculation technique as it would take weeks or months to calculate the BSPMs of all different combinations. In contrast to that, the PCA-based method needed 40 s to reconstruct 3300 signals. The conductivities that were predicted by the optimization procedure were close to the reference conductivities. If the reference signals had conductivity variations by $\pm25\%$ from the GG conductivity in all four tissues, the average deviation (over all four tissues) of the optimized conductivities was 2.96 ± 2.24 percentage points for the atria and 4.74 ± 4.21 percentage points for the ventricles.

It should be noted, that the optimization procedure needed a significant slope of the polynomial interpolation function $q(\sigma)$ (see Fig. 7.19). Otherwise, small errors in the PCA scores (optimization result) could lead to large errors in the predicted tissue conductivities. Although the slope of $q(\sigma)$ was sufficient to determine the most likely conductivities for the four important tissues under investigation, it might not be possible to use this method for organs like the spleen or kidneys, which are known to have little impact on the BSPM (and thus small slopes of $q(\sigma)$) [15].

In addition to that, the conductivity optimization procedure was based on the assumption that all BSPM changes were caused by altered tissue conductivities. Yet in a clinical setting, it is highly likely that there are other model parameters that are unknown (e.g. distribution of heterogeneous electrophysiological properties or exact sequence of myocardial activation). It is possible that changes in these unknown parameters have similar effects as changes that are caused by conductivity variations. In this case the other unknown parameters would disguise the effects of the conductivity changes and thus prevent a successful conductivity optimization. On the other hand, the PCA-based optimization method could potentially be used in these other modeling domains as well. For example it might be possible to use it for the determination of heterogeneous ion channel distributions that are important for a realistic model of ventricular repolarization.

In general, it might also be possible to apply the PCA-based technique to the inverse problem of electrocardiography (in this case, the cardiac sources should be found for a given BSPM distribution). Within this realm, the cardiac sources would have to be reconstructed for a measured BSPM several times with different conductivities. The resulting data could then be fed into the PCA-based method.

In conclusion, the presented PCA-based approach can be used to efficiently predict BSPM changes that were associated with tissue conductivity variations. BSPM changes in a single tissue were captured by the first PCA eigenvector. Signal changes due to variations in multiple tissues could be described by combining the eigenvectors from PCAs of the corresponding tissues. With this information, confidence intervals for arbitrary conductivity uncertainties within the initial conductivity boundaries could be calculated.

The eigenvectors and PCA scores are uniquely calculated for every torso model and every tissue. Thus, these parameters are patient-specific and consider individual differences in thoracic anatomy (e.g. organ size and position), cardiac anatomy and electrophysiology. The method can therefore be applied to a great range of datasets (e.g. patient-specific setups that result from anatomical and electrophysiological personalization). Based on these features, it can be concluded that the presented PCA-based approach is a promising tool to evaluate the effects of conductivity uncertainties on the outcome of clinically relevant patient-specific forward calculations.

8

Conclusions and Outlook

During the course of this thesis, several important aspects on various levels of the multiscale cardiac modeling loop have been investigated:

On the cellular level, a model for the intracellular beta-adrenergic signaling pathway was integrated into an electrophysiological model for human ventricular myocytes. The propensity of this new model towards the development of calcium sparks (which can trigger EADs) was evaluated and the model was expanded by including adrenergic regulation on I_{NaK} which allowed a more realistic description of the intracellular calcium handling under sympathetic influence. In this work, the main reason for the inclusion of the effects of beta-adrenergic regulation was its prominent role with respect to the congenital Long-QT syndrome (LQTS). In two studies, both transmural and body surface ECGs have been calculated in the absence and presence of a LQTS subtype. Although the model became more realistic if adrenergic effects were considered (especially concerning the changes in DOR), there are still some unanswered questions when it comes to the shape and amplitude of the T-Waves that are associated with LQT 1 and LQT 2.

It should also be noted, that the importance of beta-adrenergic regulation is not limited to the modeling of the LQTS. As a matter of fact, there are numerous scenarios which would benefit from the consideration of these effects (i.e. simulations at elevated heart rate or mechanical modeling due to changes in the intracellular calcium handling which affect lusitropy and inotropy).

On an organ level, we developed and evaluated a semi-automatic approach that allowed to consider the effects of the specialized excitation conduction system (with a focus on the Purkinje fiber network) during physiological ventricular activation based on a so-called endocardial stimulation profile. The presented approach delivered realistic activation sequences which were validated using standard isochrone maps and proband-specific ECG data. The semi-automatic nature of the method facilitates the fast creation of endocardial stimulation profiles on new, (patient-specific) anatomical models. A current limitation of the approach is related to the fact that the adaptation of the stimulation profile to a new anatomy or ECG still necessitates manual interaction, which can be tedious and time-consuming. In a future model version, we envision the use of an automated parameter adaptation based on a numerical optimization scheme and an adequate optimization criterion (i.e. the maximization of the correlation coefficient between simulated and measured QRS complexes). To reduce simulation times for each iteration of the parameter adaptation, a rule-based cellular automaton (for the simulation of excitation spread) in conjunction with a lead field matrix approach (for the solution of the forward problem) could be used. This should be precise enough for the simulation of ventricular activation as electrotonic coupling is here not as important as e.g. during ventricular repolarization. Using this proposed framework, one iteration (which encompasses the creation of the stimulation profile, the simulation of ventricular activation and the forward calculation of the ECG) could be performed in ≈ 5 min. This should be fast enough to optimize the parameters of the endocardial stimulation profile in a trial-and-error manner. The first steps to implement this optimization framework have already been undertaken.

Another aspect that was investigated on the level of the ventricles was related to the sequence of repolarization and the genesis of the T-Wave. A literature survey was conducted in order to create representative, heterogeneous distributions of I_{Ks} which is thought to be mainly responsible for the dispersion of APD in human ventricular tissue. Although no distribution was able to completely reproduce the repolarization pattern that was seen in a reference measurement (multi-channel ECG), it was obvious that some distributions were more likely to produce realistic results than others; i.e.:

- If M cells are considered, they should be closer to the endo- than to the epicardial wall to allow for a concordant T-Wave.
- With respect to apico-basal gradients, our simulations showed that a higher apical density of I_{Ks} (which could be translated into shorter apical APD) is more likely than a higher basal density.
- Interventricular heterogeneities did not improve the T-Wave morphology in the simulations. They rather led to biphasic T-Waves which were not seen in the reference measurements.

On the torso level, an investigation evaluated the realism of different (simple) rule-based approaches to model the skeletal muscle fiber orientation. Although one of the approaches was published in prestigious journal [243] we were surprised that none of the rule-based methods was able to represent the skeletal muscle fiber anisotropy appropriately. For future studies, there are two possibilities: more complex rule-based approaches could be created and tested or efforts could be made to image the skeletal muscle fiber orientation in-vivo using DTMRI sequences.

In the realm of the solution of the forward problem, we investigated the impact of ventricular deformation and the associated movement of the electrical sources on the morphology of the T-Wave in the ECG. Among the new aspects of this study was the fact that the deformation was based on MR image data and that the simulations were conducted using a 3D model of the ventricles rather than a 2 D slice. The main effect on the T-Wave was a reduction of its amplitude which was both plausible and in agreement with other studies. As the approach chosen in this study relied heavily on the manual segmentation of the ventricles in differently contracted (and relaxed) states, it is not possible to use it regularly whenever new patient-specific models become available. Two strategies can be used to make this limitation tolerable: on the one hand it is possible to at least roughly consider the effects of deformation by simply scaling the simulated T-Waves with the factors determined in this study (to consider the amplitude reduction). On the other hand, we proposed a possibility to enhance the generation of the dynamic heart and torso model which eliminates the need for manual segmentation. In this case, the movement is no longer captured by cinematographic MRI scans but rather by a large number of tagging slices. From these tagging data, it is possible to automatically extract the displacement information based on the so-called Sine Wave Modeling approach [391]. Thus, the registration procedure and subsequently the creation of the dynamic model could be completely automated.

Finally, the effects of tissue conductivities on the solution of the forward problem were investigated. In a first study, we conducted a *sensitivity* analysis to evaluate the most important inhomogeneities for a realistic solution of the forward problem. In addition to that, an *uncertainty* analysis evaluated the effects of contradictory conductivity reports from different experimental studies. Finally we removed groups of organs from the torso model and calculated the associated errors with respect to a reference model. This allowed to propose recommendations which organs have to be considered during the creation of patient-specific models. Among the five most important structures (besides the heart) were: skeletal muscle conductivity as well as anisotropy, blood, lungs, and fat.

In a second study that dealt with the role of tissue conductivities, we developed a new and efficient method that allowed to predict the effects of tissue conductivity changes on the BSPM. This method was based on the principal component analysis and was significantly faster than repetitive forward calculation which represented the only option that was available to investigate these effects so far. Thus, the proposed approach can potentially be used in the future to personalize in-silico models in a clinical setting.

In conclusion, it can be said that although there are still numerous unanswered questions in the area of multiscale cardiac modeling. Yet its potential to support clinical diagnosis and therapy of cardiovascular diseases is undisputed. In this context, the presented thesis has advanced the state-of-the-art in several important areas. The associated increase in model realism and personalization is one of many steps towards the clinical use of quantitative in-silico modeling of the heart.

References

1. R. Kalayciyan, "Einfluss des endokardialen Stimulationsprofils auf die Morphologie des Oberflächen-EKGs," Master's thesis, 2009.

2. D. U. J. Keller, R. Kalayciyan, O. Dössel, and G. Seemann, "Fast creation of endocardial stimulation profiles for the realistic simulation of body surface ECGs," in *Proc. IFMBE*, vol. 25/4, pp. 145–148, 2009.

3. R. Kalayciyan, D. U. J. Keller, G. Seemann, and O. Dössel, "Creation of a realistic endocardial stimulation profile for the visible man dataset," in *Proc. IFMBE*, vol. 25/4, pp. 934–937, 2009.

4. D. U. J. Keller, D. L. Weiss, O. Dössel, and G. Seemann, "Influence of I_{Ks} Heterogeneities on the Genesis of the T-Wave: A Computational Evaluation," *IEEE Trans. Biomed. Eng.*, 2011 (submitted).

5. C. A. Otto, "Impact of adrenergic regulation on the electrophysiological properties of human ventricular myocytes," Master's thesis, 2009.

6. A. Bohn, "Einfluss β-adrenerger Regulation auf die Elektrophysiologie und Repolarisation im menschlichen Ventrikel," Master's thesis, 2009.

7. C. A. Otto, D. U. J. Keller, G. Seemann, and O. Dössel, "Integrating Beta-Adrenergic Signaling into a Computational Model of Human Cardiac Electrophysiology," in *Proc. IFMBE*, vol. 25/4, pp. 1033–1036, 2009.

8. D. U. J. Keller, G. Seemann, D. L. Weiss, D. Farina, J. Zehelein, and O. Dössel, "Computer based modeling of the congenital long-QT 2 syndrome in the Visible Man torso: From genes to ECG," in *Conf. Proc. IEEE Eng. Med. Biol. Soc.*, pp. 1410–1413, 2007.

9. D. U. J. Keller, D. L. Weiss, O. Dössel, and G. Seemann, "Evaluating mathematical cell models as to their suitability for modeling the congenital long-QT 2 syndrome," in *Proc. BMT*, vol. 52, 2007.

10. D. U. J. Keller, A. Bohn, O. Dössel, and S. G., "In-silico Evaluation of Beta-Adrenergic Effects on the Long-QT Syndrome," in *Proc. CinC*, vol. 37, pp. 825–828, 2010.

11. D. U. J. Keller, O. Dössel, and G. Seemann, "Simulating cardiac excitation in a high resolution biventricular model," in *Proc. BMT*, vol. Supplement 2010, 2010.

12. D. U. J. Keller, O. Dössel, and G. Seemann, "Evaluation of Rule-Based Approaches for the Incorporation of Skeletal Muscle Fiber Orientation in Patient-Specific Anatomies," in *Proc. CinC*, vol. 36, pp. 181–184, 2009.

13. D. U. J. Keller, O. Jarrousse, T. Fritz, S. Ley, O. Dössel, and G. Seemann, "Impact of Physiological Ventricular Deformation on the Morphology of the T-Wave: A Hybrid, Static-Dynamic Approach," *IEEE Trans. Biomed. Eng.*, 2011 (accepted).

14. P. Tri Dung, "The impact of distinctive organ conductivities on the solution of the forward problem of electrocardiography," Master's thesis, 2008.

15. D. U. J. Keller, F. M. Weber, G. Seemann, and O. Dössel, "Ranking the influence of tissue conductivities on forward-calculated ECGs," *IEEE Trans. Biomed. Eng.*, vol. 57, pp. 1568–1576, 2010.

16. S. Bauer, "Efficient reconstruction of BSPMs and optimization of multiple tissue conductivities," Master's thesis, 2009.

17. S. Bauer, D. U. J. Keller, F. M. Weber, P. Tri Dung, G. Seemann, and O. Dössel, "How do tissue conductivities impact on forward-calculated ECGs? An efficient prediction based on principal component analysis," in *Proc. IFMBE*, vol. 25/4, pp. 641–644, 2009.

18. F. M. Weber, D. U. J. Keller, S. Bauer, G. Seemann, C. Lorenz, and O. Dössel, "Predicting Tissue Conductivity Influences on Body Surface Potentials - An Efficient Approach based on Principal Component Analysis," *IEEE Trans. Biomed. Eng.*, vol. 58, pp. 265–273, 2011.

19. "Kellogg-Academy." http://academic.kellogg.cc.mi.us.

20. F. H. Netter, *Atlas der Anatomie des Menschen*. Basel: Ciba-Geigy AG, 1995.

21. F. H. Martini, *Fundamentals of Anatomy & Physiology*. Prentice Hall, Inc. Upper Saddle River, NJ: Benjamin-Cummings Publishing Company, 2006.

22. K. Saladin, *Anatomy and Physiology : The Unity of Form and Function*. McGraw-Hill, 2003.

23. C. Henriquez, A. Muzikant, and C. Smoak, "Anisotropy, Fiber Curvature and Bath Loading Effects on Activation in Thin and Thick Cardiac Tissue Preparations: Simulations in a Three-Dimensional Bidomain Model," *J. Cardiovasc. Electrophysiol.*, vol. 7, pp. 424–444, 1996.

24. J. Malmivuo and R. Plonsey, *Bioelectromagnetism*, ch. Bidomain Model of Multicellular Volume Conductors, pp. 159–168. New York; Oxford: Oxford University Press, 1995.

25. P. S. Jouk, Y. Usson, G. Michalowicz, and L. Grossi, "Three-dimensional cartography of the pattern of the myofibres in the second trimester fetal human heart," *Anat. Embryol.*, vol. 202, pp. 103–118, 2000.

26. P. Nielsen, I. LeGrice, P. Hunter, and B. Smaill, "Mathematical Model of Geometry and Fibrous Structure of the Heart," *Am. J. Physiol.*, vol. 260, pp. H1365–H1378, 1991.

27. P. P. Lunkenheimer, K. Redmann, N. Kling, X. Jiang, K. Rothaus, C. W. Cryer, F. Wubbeling, P. Niederer, P. U. Heitz, S. Y. Ho, and R. H. Anderson, "Three-dimensional architecture of the left ventricular myocardium," *Anat. Rec. A Discov. Mol. Cell. Evol. Biol.*, vol. 288, pp. 565–578, 2006.

28. L. Zhukov and A. Barr, "Heart-muscle fiber reconstruction from diffusion tensor MRI," in *Proceedings of IEEE Visualization*, pp. 597–602, 2003.

29. L. Geerts, P. Bovendeerd, K. Nicolay, and T. Arts, "Characterization of the normal cardiac myofiber field in goat measured with MR-diffusion tensor imaging," *Am. J. Physiol. Heart. Circ. Physiol.*, vol. 283, p. 139, 2002.

30. D. F. Scollan, A. A. Holmes, J. Zhang, and R. L. Winslow, "Reconstruction of myocardial architecture at high resolution using diffusion tensor MRI," in *Conf. Proc. IEEE Eng. Med. Biol. Soc.*, p. 1071, 1999.

31. D. F. Scollan, A. Holmes, J. Zhang, and R. L. Winslow, "Reconstruction of cardiac ventricular geometry and fiber orientation using magnetic resonance imaging," *Ann. Biomed. Eng.*, vol. 28, pp. 934–944, 2000.

32. D. Rohmer, A. Sitek, and G. T. Gullberg, "Reconstruction and visualization of fiber and laminar structure in the normal human heart from ex vivo diffusion tensor magnetic resonance imaging (DTMRI)

data," *Invest. Radiol.*, vol. 42, pp. 777–789, 2007.

33. P. Helm, M. F. Beg, M. I. Miller, and R. L. Winslow, "Measuring and mapping cardiac fiber and laminar architecture using diffusion tensor MR imaging," *Ann. NY Acad. Sci.*, vol. 1047, pp. 296–307, 2005.

34. P. Bovendeerd, J. Rijcken, D. H. V. Campen, A. J. G. Schoofs, K. Nicolay, and T. Arts, "Optimization of left ventricular muscle fiber orientation," *IUTAM Symposium on Synthesis in Bio Solid Mechanics*, pp. 285–296, 1999.

35. D. Streeter, *Handbook of Physiology. Section 2: The Cardiovascular System*, vol. Vol. I. The Heart, ch. Gross morphology and fiber geometry of the heart, pp. 61–112. Baltimore: Williams and Wilkins: American Physiology Society, 1979.

36. J. Rijcken, P. H. M. Bovendeerd, A. J. G. Schoofs, D. H. vanCampen, and T. Arts, "Optimization of cardiac fiber orientation for homogeneous fiber strain during ejection," *Ann. Biomed. Eng.*, vol. 27, pp. 289–297, 1999.

37. D. U. J. Keller, "Detailed anatomical and electrophysiological modeling of human ventricles based on diffusion tensor MRI," Master's thesis, 2006.

38. D. L. Weiß, "Charakterisierung der Ventrikelwand durch anatomische und elektrophysiologische Modellierung," Master's thesis, 2003.

39. I. J. LeGrice, B. H. Smaill, L. Z. Chai, S. G. Edgar, J. B. Gavin, and P. J. Hunter, "Laminar structure of the heart: ventricle myocyte arrangement and connective tissue architecture in the dog," *Am. J. Physiol.*, vol. 269, pp. H571–82, 1995.

40. D. D. Streeter, H. M. Spotnitz, D. P. Patel, R. J. J.R., and E. Sonnenblick, "Fiber orientation in the canine left ventricle during diastole and systole," *Circ. Res.*, vol. 24, pp. 339–347, 1969.

41. J. M. Rijcken, *Optimization of left ventricular fiber orientation*. PhD thesis, 1997.

42. D. F. Scollan, *Reconstructing the heart: Development and application of biophysically-based electrical models of propagation in ventricular myocardium reconstructed from diffusion tensor mri*. PhD thesis, 2002.

43. G. Seemann, D. U. J. Keller, D. L. Weiss, and O. Dössel, "Modeling Human Ventricular Geometry and Fiber Orientation Based on Diffusion Tensor MRI," in *Proc. CinC*, vol. 33, pp. 801–804, 2006.

44. R. H. Anderson, S. Y. Ho, D. Sanchez-Quintana, K. Redmann, and P. P. Lunkenheimer, "Heuristic problems in defining the three-dimensional arrangement of the ventricular myocytes," *Anat. Rec. A Discov. Mol. Cell. Evol. Biol.*, vol. 288, pp. 579–586, 2006.

45. I. J. LeGrice, P. J. Hunter, and B. H. Smaill, "Laminar structure of the heart: a mathematical model," *Am. J. Physiol.*, vol. 272, pp. H2466–H2476, 1997.

46. G. Seemann, *Modeling of electrophysiology and tension development in the human heart*. PhD thesis, Universitätsverlag Karlsruhe, 2005.

47. R. F. Schmidt, *Physiologie kompakt*, ch. Erregungsphysiologie des Herzens, pp. 197–204. Berlin, Heidelberg, New York: Springer, 1999.

48. S. H. Litkovsky and C. Antzelevitch, "Rate dependence of action potential duration and refractoriness in canine ventricular endocardium differs from that of epicardium: The role of transient outward current," *J. Am. Coll. Cardiol.*, vol. 14, pp. 1053–1066, 1989.

49. A. Lukas and C. Antzelevitch, "Differences in the electrophysiological response of canine ventricular epicaridium and endocardium to ischemia: Role of the transient outward current," *Circulation*, vol. 88, pp. 2903–2915, 1993.

50. T. Furukawa, R. J. Myerburg, N. Furukawa, A. L. Bassett, and S. Kimura, "Differences in transient outward currents of feline endocardial and epicardial myocytes," *Circ. Res.*, vol. 67, pp. 1287–1291, 1990.

51. D. Fedida and W. R. Giles, "Regional variations in action potentials and transient outward current in myocytes isolated from rabbit left ventricle," *J. Physiol.*, vol. 442, pp. 191–209, 1991.

52. Y. Shimoni, D. Severson, and W. Giles, "Thyroid status and diabetes modulate regional differences in potassium currents in rat ventricle," *J. Physiol.*, vol. 488, pp. 673–688, 1995.

53. D. W. Liu, G. A. Gintant, and C. Antzelevitch, "Ionic bases for electrophysiological distinctions among epicardial, midmyocardial, and endocardial myocytes from the free wall of the canine left ventricle," *Circ. Res.*, vol. 72, pp. 671–687, 1993.

54. G. R. Li, J. Feng, L. Yue, and M. Carrier, "Transmural heterogeneity of action potentials and I_{to1} in myocytes isolated from the human right ventricle," *Am. J. Physiol.*, vol. 275, pp. 369–377, 1998.

55. M. Näbauer, D. J. Beuckelmann, P. Überfuhr, and G. Steinbeck, "Regional differences in current density and rate-dependent properties of the transient outward current in subepicardial and subendocardial myocytes of human left ventricle," *Circulation*, vol. 93, pp. 168–177, 1996.

56. J. M. D. Diego, Z. Q. Sun, and C. Antzelevitch, "I(to) and action potential notch are smaller in left vs. right canine ventricular epicardium," *Am. J. Physiol.*, vol. 271, pp. H548–H561, 1996.

57. P. G. Volders, K. R. Sipido, E. Carmeliet, R. L. Spatjens, H. J. Wellens, and M. A. Vos, "Repolarizing K^+ currents ITO1 and IKs are larger in right than left canine ventricular midmyocardium," *Circulation*, vol. 99, pp. 206–210, 1999.

58. T. Furukawa, S. Kimura, N. Furukawa, A. L. Bassett, and R. J. Myerburg, "Potassium rectifier current differ in myocytes of endocardial and epicardial origin," *Circ. Res.*, vol. 70, pp. 91–103, 1992.

59. Z. Wang, L. Yue, M. White, G. Pelletier, and S. Nattel, "Differential distribution of inward rectifier potassium channel transcripts in human atrium versus ventricle," *Circulation*, vol. 98, pp. 2422–2428, 1998.

60. A. C. Zygmunt, R. J. Goodrow, and C. Antzelevitch, "I(NaCa) contributes to electrical heterogeneity within the canine ventricle," *Am. J. Physiol.*, vol. 278, pp. H1671–H1678, 2000.

61. A. C. Zygmunt, G. T. Eddlestone, G. P. Thomas, V. V. Nesterenko, and C. Antzelevitch, "Larger late sodium conductance in M Cells contributes to electrical heterogeneity in canine ventricle," *Am. J. Physiol.*, vol. 281, pp. H689–H697, 2001.

62. K. R. Laurita, R. Katra, B. Wible, X. Wan, and M. H. Koo, "Transmural heterogeneity of calcium handling in canine," *Circ. Res.*, vol. 92, pp. 668–675, 2003.

63. C. Antzelevitch, S. Sicouri, S. H. Litovsky, A. Lukas, S. C. Krishman, J. M. D. Diego, G. A. Gintant, and D. W. Liu, "Heterogeneity within the ventricular wall: electrophysiology and pharmacology of epicardial, endocardial, and M-cells," *Circ. Res.*, vol. 69, pp. 1427–1449, 1991.

64. J. M. Cordeiro, L. Greene, C. Heilmann, D. Antzelevitch, and C. Antzelevitch, "Transmural heterogeneity of calcium activity and mechanical function in the canine left ventricle," *Am. J. Physiol.*, vol. 286, pp. H1471–H1479, 2003.

65. E. Hermeling, T. M. Verhagen, F. W. Prinzen, and N. H. L. Kuijpers, "Transmural heterogeneity in ion channel properties in the left ventricle optimizes pump function during natural electrical activation," in *Proc. CinC*, vol. 36, pp. 397–400, 2009.

66. Y. Pereon, S. Demolombe, I. Baro, E. Drouin, F. Charpentier, and D. Escande, "Differential expression of KvLQT1 isoforms across the human ventricular wall," *Am. J. Physiol-Heart C.*, vol. 278,

p. 1908, 2000.

67. F. L. Burton and S. M. Cobbe, "Dispersion of ventricular repolarization and refractory period," *Cardiovasc. Res.*, vol. 50, pp. 10–23, 2001.

68. J. M. T. de Bakker and T. Opthof, "Is the apico-basal gradient larger than the transmural gradient?," *J. Cardiovasc. Pharm.*, vol. 39, pp. 328–331, 2002.

69. D. Noble and I. Cohen, "The interpretation of the T wave of the electrocardiogram," *Cardiovasc. Res.*, vol. 12, pp. 13–27, 1978.

70. S. Y. Ho and D. Sanchez-Quintana, "The importance of atrial structure and fibers," *Clin. Anat.*, vol. 22, pp. 52–63, 2008.

71. T. S. Lewis, *The Mechanism and Graphic Registration of the Heart Beat*, vol. Chapter 1, ch. Special Anatomy, pp. 1–14. London: Shaw & Sons Ltd., 1920.

72. D. C. Myers and G. I. Fishman, "Toward an understanding of the genetics of murine cardiac pacemaking and conduction system development," *Anat. Rec. Part A*, vol. 280, pp. 1018–1021, 2004.

73. L. Miquerol, S. Meysen, M. Mangoni, P. Bois, H. V. M. van Rijen, P. Abran, H. Jongsma, J. Nargeot, and D. Gros, "Architectural and functional asymmetry of the His-Purkinje system of the murine heart," *Cardiovasc. Res.*, vol. 63, pp. 77–86, 2004.

74. H. Tanaka, T. Hamamoto, and T. Takamatsu, "Toward an Integrated Understanding of the Purkinje Fibers in the Heart: The Functional and Morphological Interconnection between the Purkinje Fibers and Ventricular Muscle," *Acta. Histochemica et Cytochemica*, vol. 38, pp. 257–265, 2005.

75. S. Tawara, *Das Reitzleitungssystem des Säugetierherzen*. Jena: G. Fischer, 1906.

76. W. Han, W. Bao, Z. Wang, and S. Nattel, "Comparison of ion-channel subunit expression in canine cardiac Purkinje fibers and ventricular muscle," *Circ. Res.*, vol. 91, pp. 790–797, 2002.

77. J. M. Cordeiro, K. W. Spitzer, and W. R. Giles, "Repolarizing K+ currents in rabbit heart Purkinje cells," *The Journal of Physiology*, vol. 508 (Pt 3), pp. 811–823, 1998.

78. G. Schram, M. Pourrier, P. Melnyk, and S. Nattel, "Differential distribution of cardiac ion channel expression as a basis for regional specialization in electrical function," *Circ. Res.*, vol. 90, pp. 939–950, 2002.

79. B. Balati, A. Varro, and J. G. Papp, "Comparison of the cellular electrophysiological characteristics of canine left ventricular epicardium, M cells, endocardium and Purkinje fibres," *Acta Physiol Scand*, vol. 164, pp. 181–190, 1998.

80. H. R. Lu, R. Marien, A. Saels, and F. De Clerck, "Species plays an important role in drug-induced prolongation of action potential duration and early afterdepolarizations in isolated Purkinje fibers," *J. Cardiovasc. Electrophysiol.*, vol. 12, pp. 93–102, 2001.

81. T. N. James, "Anatomy of the cardiac conduction system in the rabbit," *Circ. Res.*, vol. 20, pp. 638–648, 1967.

82. P. W. Oosthoek, S. Viragh, W. H. Lamers, and A. F. Moorman, "Immunohistochemical delineation of the conduction system. II: The atrioventricular node and Purkinje fibers," *Circ. Res.*, vol. 73, pp. 482–491, 1993.

83. G. K. Massing and T. N. James, "Anatomical configuration of the His bundle and bundle branches in the human heart," *Circulation*, vol. 53, pp. 609–621, 1976.

84. M. B. Rosenbaum, M. V. Elizari, and J. O. Lazzari, *The Hemiblocks. New Concepts of Intraventricular Conduction Based on Human Anatomical, Physiological and Clinical Studies*. Tampa: Oldsmar, FL, 1970.

85. H. N. Uhley, "The concept of trifascicular intraventricular conduction: historical aspects and influence on contemporary cardiology," *Am. J. Cardiol.*, vol. 43, pp. 643–646, 1979.

86. J. C. Demoulin and H. E. Kulbertus, "Histopathological examination of concept of left hemiblock," *Brit. Heart J.*, vol. 34, pp. 807–814, 1972.

87. S. Silbernagl and A. Despopoulos, *Taschenatlas der Physiologie.* Stuttgart: Thieme, 2006.

88. T. Pollard and W. Earnshaw, *Cell Biology.* Saunders, 2 ed., 2007.

89. K. Piippo, H. Swan, M. Pasternack, H. Chapman, K. Paavonen, M. Viitasalo, L. Toivonen, and K. Kontula, "A founder mutation of the potassium channel KCNQ1 in long QT syndrome: implications for estimation of disease prevalence and molecular diagnostics," *J. Am. Coll. Cardiol.*, vol. 37, pp. 562–568, 2001.

90. M. J. Ackerman, "Cardiac channelopathies: it's in the genes," *Nature Medicine*, vol. 10, pp. 463–464, 2004.

91. G. M. Vincent, "The long-QT syndrome–bedside to bench to bedside," *N. Engl. J. Med.*, vol. 348, pp. 1837–1838, 2003.

92. A. J. Moss and P. J. Schwartz, "25th anniversary of the International Long-QT Syndrome Registry: an ongoing quest to uncover the secrets of long-QT syndrome," *Circulation*, vol. 111, pp. 1199–1201, 2005.

93. G. M. Vincent, K. W. Timothy, M. Leppert, and M. Keating, "The spectrum of symptoms and QT intervals in carriers of the gene for the long-QT syndrome," *N. Engl. J. Med.*, vol. 327, pp. 846–852, 1992.

94. P. J. Schwartz, A. J. Moss, G. M. Vincent, and R. S. Crampton, "Diagnostic criteria for the long QT syndrome. An update," *Circulation*, vol. 88, pp. 782–784, 1993.

95. T. Lewalter, S. Englisch, S. Jahr, B. Schumacher, W. Jung, R. D. Hesch, and B. Luederitz, "QT-Syndrome: Neue diagnostische Moeglichkeiten," *Zeitschrift fuer Kardiologie*, vol. 87, pp. 517–521, 1998.

96. R. S. Kass and A. J. Moss, "Long QT syndrome: novel insights into the mechanisms of cardiac arrhythmias," *J. Clin. Invest.*, vol. 112, pp. 810–815, 2003.

97. W. Zareba and I. Cygankiewicz, "Long QT syndrome and short QT syndrome," *Prog. Cardiovasc. Dis.*, vol. 51, pp. 264–278, 2008.

98. A. Medeiros-Domingo, P. Iturralde-Torres, and M. J. Ackerman, "Clinical and genetic characteristics of long QT syndrome," *Rev. Esp. Cardiol.*, vol. 60, pp. 739–752, 2007.

99. L. Chen, M. L. Marquardt, D. J. Tester, K. J. Sampson, M. J. Ackerman, and R. S. Kass, "Mutation of an A-kinase-anchoring protein causes long-QT syndrome," *P. Natl. Acad. Sci. USA*, vol. 104, pp. 20990–20995, 2007.

100. J. J. Saucerman, S. N. Healy, M. E. Belik, J. L. Puglisi, and A. D. McCulloch, "Proarrhythmic consequences of a KCNQ1 AKAP-binding domain mutation: computational models of whole cells and heterogeneous tissue," *Circ. Res.*, vol. 95, pp. 1216–1224, 2004.

101. I. M. Van Langen, E. Birnie, M. Alders, R. J. Jongbloed, H. Le Marec, and A. A. M. Wilde, "The use of genotype-phenotype correlations in mutation analysis for the long QT syndrome," *J. Med. Genet.*, vol. 40, pp. 141–145, 2003.

102. P. J. Schwartz, S. G. Priori, C. Spazzolini, A. J. Moss, G. M. Vincent, C. Napolitano, I. Denjoy, P. Guicheney, G. Breithardt, M. T. Keating, J. A. Towbin, A. H. Beggs, P. Brink, A. A. Wilde, L. Toivonen, W. Zareba, J. L. Robinson, K. W. Timothy, V. Corfield, D. Wattanasirichaigoon, C. Cor-

bett, W. Haverkamp, E. Schulze-Bahr, M. H. Lehmann, K. Schwartz, P. Coumel, and R. Bloise, "Genotype-phenotype correlation in the long-QT syndrome: gene-specific triggers for life-threatening arrhythmias," *Circulation*, vol. 103, pp. 89–95, 2001.

103. I. Goldenberg and A. J. Moss, "Long QT syndrome," *J. Am. Coll. Cardiol.*, vol. 51, pp. 2291–2300, 2008.

104. P. J. Schwartz and A. Malliani, "Electrical alternation of the T-wave: clinical and experimental evidence of its relationship with the sympathetic nervous system and with the long Q-T syndrome," *Am. Heart J.*, vol. 89, pp. 45–50, 1975.

105. T. Noda, H. Takaki, T. Kurita, K. Suyama, N. Nagaya, A. Taguchi, N. Aihara, S. Kamakura, K. Sunagawa, K. Nakamura, T. Ohe, M. Horie, C. Napolitano, J. A. Towbin, S. G. Priori, Shimizu, and W., "Gene-specific response of dynamic ventricular repolarization to sympathetic stimulation in LQT1, LQT2 and LQT3 forms of congenital long QT syndrome," *Eur. Heart J.*, vol. 23, pp. 975–983, 2002.

106. P. J. Schwartz, E. Vanoli, L. Crotti, C. Spazzolini, C. Ferrandi, A. Goosen, P. Hedley, M. Heradien, S. Bacchini, A. Turco, M. T. La Rovere, A. Bartoli, A. L. J. George, and P. A. Brink, "Neural control of heart rate is an arrhythmia risk modifier in long QT syndrome," *J. Am. Coll. Cardiol.*, vol. 51, pp. 920–929, 2008.

107. S. O. Marx, J. Kurokawa, S. Reiken, H. Motoike, J. D'Armiento, A. R. Marks, and R. S. Kass, "Requirement of a macromolecular signaling complex for beta adrenergic receptor modulation of the KCNQ1-KCNE1 potassium channel," *Science*, pp. 496–499, 2002.

108. R. S. Kass, J. Kurokawa, S. O. Marx, and A. R. Marks, "Leucine/isoleucine zipper coordination of ion channel macromolecular signaling complexes in the heart. Roles in inherited arrhythmias," *Trends Cardiocas. Med.*, vol. 13, pp. 52–56, 2003.

109. C. Terrenoire, C. E. Clancy, J. W. Cormier, K. J. Sampson, and R. S. Kass, "Autonomic control of cardiac action potentials: role of potassium channel kinetics in response to sympathetic stimulation," *Circ. Res.*, vol. 96, pp. e25–34, 2005.

110. C. Antzelevitch, "Sympathetic modulation of the long QT syndrome," *Eur. Heart J.*, vol. 23, pp. 1246–1252, 2002.

111. C. E. Conrath and T. Opthof, "Ventricular repolarization: an overview of (patho)physiology, sympathetic effects and genetic aspects," *Prog. Biophys. Mol. Bio.*, vol. 92, pp. 269–307, 2006.

112. S. G. Priori, D. W. Mortara, C. Napolitano, L. Diehl, V. Paganini, F. Cantu, G. Cantu, and P. J. Schwartz, "Evaluation of the spatial aspects of T-wave complexity in the long-QT syndrome," *Circulation*, vol. 96, pp. 3006–3012, 1997.

113. L. De Ambroggi, T. Bertoni, E. Locati, M. Stramba-Badiale, and P. J. Schwartz, "Mapping of body surface potentials in patients with the idiopathic long QT syndrome," *Circulation*, vol. 74, pp. 1334–1345, 1986.

114. L. Zhang, K. W. Timothy, G. M. Vincent, M. H. Lehmann, J. Fox, L. C. Giuli, J. Shen, I. Splawski, S. G. Priori, S. J. Compton, F. Yanowitz, J. Benhorin, A. J. Moss, P. J. Schwartz, J. L. Robinson, Q. Wang, W. Zareba, M. T. Keating, J. A. Towbin, C. Napolitano, and A. Medina, "Spectrum of ST-T-wave patterns and repolarization parameters in congenital long-QT syndrome: ECG findings identify genotypes," *Circulation*, vol. 102, pp. 2849–2855, 2000.

115. W. Zareba, "Genotype-specific ECG patterns in long QT syndrome," *J. Electrocardiol.*, vol. 39, pp. S101–6, 2006.

116. J. M. Lupoglazoff, I. Denjoy, M. Berthet, N. Neyroud, L. Demay, P. Richard, B. Hainque, G. Vaksmann, D. Klug, A. Leenhardt, G. Maillard, P. Coumel, and P. Guicheney, "Notched T waves on Holter recordings enhance detection of patients with LQT2 (HERG) mutations," *Circulation*, vol. 103, pp. 1095–1101, 2001.

117. K. Takenaka, T. Ai, W. Shimizu, A. Kobori, T. Ninomiya, H. Otani, T. Kubota, H. Takaki, S. Kamakura, and M. Horie, "Exercise stress test amplifies genotype-phenotype correlation in the LQT1 and LQT2 forms of the long-QT syndrome," *Circulation*, vol. 107, pp. 838–844, 2003.

118. A. J. Moss, W. Zareba, J. Benhorin, E. H. Locati, W. J. Hall, J. L. Robinson, P. J. Schwartz, J. A. Towbin, G. M. Vincent, and M. H. Lehmann, "ECG T-wave patterns in genetically distinct forms of the hereditary long QT syndrome," *Circulation*, vol. 92, pp. 2929–2934, 1995.

119. C. Antzelevitch, "Heterogeneity of cellular repolarization in LQTS: the role of M cells," *Eur. Heart J. Suppl.*, vol. 3, pp. K2–K16, 2001.

120. J. J. Struijk, J. K. Kanters, M. P. Andersen, T. Hardahl, C. Graff, M. Christiansen, and E. Toft, "Classification of the long-QT syndrome based on discriminant analysis of T-wave morphology," *Med. Biol. Eng. Comput.*, vol. 44, pp. 543–549, 2006.

121. K. Hashiba, "Sex differences in phenotypic manifestation and gene transmission in the Romano-Ward syndrome," *Ann. NY Acad. Sci.*, vol. 644, pp. 142–156, 1992.

122. M. H. Lehmann, K. W. Timothy, D. Frankovich, B. S. Fromm, M. Keating, E. H. Locati, R. T. Taggart, J. A. Towbin, A. J. Moss, P. J. Schwartz, and G. M. Vincent, "Age-gender influence on the rate-corrected QT interval and the QT-heart rate relation in families with genotypically characterized long QT syndrome," *J. Am. Coll. Cardiol.*, vol. 29, pp. 93–99, 1997.

123. E. H. Locati, W. Zareba, A. J. Moss, P. J. Schwartz, G. M. Vincent, M. H. Lehmann, J. A. Towbin, S. G. Priori, C. Napolitano, J. L. Robinson, M. Andrews, K. Timothy, and W. J. Hall, "Age- and sex-related differences in clinical manifestations in patients with congenital long-QT syndrome: findings from the International LQTS Registry," *Circulation*, vol. 97, pp. 2237–2244, 1998.

124. C. E. Conrath, A. A. M. Wilde, R. J. E. Jongbloed, M. Alders, I. M. van Langen, J. P. van Tintelen, P. A. Doevendans, and T. Opthof, "Gender differences in the long QT syndrome: effects of beta-adrenoceptor blockade," *Cardiovasc. Res.*, vol. 53, pp. 770–776, 2002.

125. W. Zareba, A. J. Moss, J. P. Daubert, W. J. Hall, J. L. Robinson, and M. Andrews, "Implantable cardioverter defibrillator in high-risk long QT syndrome patients," *J. Cardiovasc. Electr.*, vol. 14, pp. 337–341, 2003.

126. S. G. Priori, C. Napolitano, P. J. Schwartz, M. Grillo, R. Bloise, E. Ronchetti, C. Moncalvo, C. Tulipani, A. Veia, G. Bottelli, and J. Nastoli, "Association of long QT syndrome loci and cardiac events among patients treated with beta-blockers," *JAMA-J. Am. Med. Assoc.*, vol. 292, pp. 1341–1344, 2004.

127. S. J. Compton, R. L. Lux, M. R. Ramsey, K. R. Strelich, M. C. Sanguinetti, L. S. Green, M. T. Keating, and J. W. Mason, "Genetically defined therapy of inherited long-QT syndrome. Correction of abnormal repolarization by potassium," *Circulation*, vol. 94, pp. 1018–1022, 1996.

128. P. J. Schwartz, S. G. Priori, E. H. Locati, C. Napolitano, F. Cantu, J. A. Towbin, M. T. Keating, H. Hammoude, A. M. Brown, and L. S. Chen, "Long QT syndrome patients with mutations of the SCN5A and HERG genes have differential responses to Na+ channel blockade and to increases in heart rate. Implications for gene-specific therapy," *Circulation*, vol. 92, pp. 3381–3386, 1995.

129. P. J. Schwartz, E. H. Locati, A. J. Moss, R. S. Crampton, R. Trazzi, and U. Ruberti, "Left cardiac sympathetic denervation in the therapy of congenital long QT syndrome. A worldwide report," *Circulation*, vol. 84, pp. 503–511, 1991.

130. C. E. Chiang and D. M. Roden, "The long QT Syndromes: genetic basis and clinical implications," *J. Am. Coll. Cardiol.*, vol. 26, pp. 1–12, 2000.

131. X. H. T. Wehrens, M. A. Vos, P. A. Doevendans, and H. J. J. Wellens, "Novel insights in the congenital long QT syndrome," *Ann. Intern. Med.*, vol. 137, pp. 981–992, 2002.

132. W. Shimizu and C. Antzelevitch, "Sodium channel block with mexiletine is effective in reducing dispersion of repolarization and preventing torsade de pointes in LQT2 and LQT3 models of the long-QT syndrome," *Circulation*, vol. 96, pp. 2038–2047, 1997.

133. K. Gima and Y. Rudy, "Ionic current basis of electrocardiographic waveforms. A model study," *Circ. Res.*, vol. 90, pp. 889–896, 2002.

134. J. W. Mason, D. J. Ramseth, D. O. Chanter, T. E. Moon, D. B. Goodman, and B. Mendzelevski, "Electrocardiographic reference ranges derived from 79,743 ambulatory subjects," *J. Electrocardiol.*, vol. 40, pp. 228–234, 2007.

135. Y. Watanabe, "Purkinje repolarization as a possible cause of the U wave in the electrocardiogram," *Circulation*, vol. 51, pp. 1030–1037, 1975.

136. B. Surawicz, "U wave: facts, hypotheses, misconceptions, and misnomers," *J. Cardiovasc. Electr.*, vol. 9, pp. 1117–1128, 1998.

137. R. Schimpf, C. Antzelevitch, D. Haghi, C. Giustetto, A. Pizzuti, F. Gaita, C. Veltmann, C. Wolpert, and M. Borggrefe, "Electromechanical coupling in patients with the short QT syndrome: further insights into the mechanoelectrical hypothesis of the U wave," *Heart Rhythm*, vol. 5, pp. 241–245, 2008.

138. H. J. Ritsema van Eck, J. A. Kors, and G. van Herpen, "The U wave in the electrocardiogram: a solution for a 100-year-old riddle," *Cardiovasc. Res.*, vol. 67, pp. 256–262, 2005.

139. C. Werner, *Simulation der elektrischen Erregungsausbreitung in anatomischen Herzmodellen mit adaptiven zellulären Automaten*. PhD thesis, 2001.

140. A. Khawaja, *Automatic ECG analysis using principal component analysis and wavelet transformation*. Karlsruhe: Universitätsverlag, 2006.

141. M. Fereniec, G. Stix, M. Kania, T. Mroczka, D. Janusek, and R. Maniewski, "Risk assessment of ventricular arrhythmia using new parameters based on high resolution body surface potential mapping," *Med. Sci. Monit.*, vol. 17, pp. Mt26–33, 2011.

142. P. Korhonen, T. Husa, T. Konttila, I. Tierala, M. Makijarvi, H. Vaananen, and L. Toivonen, "Complex T-wave morphology in body surface potential mapping in prediction of arrhythmic events in patients with acute myocardial infarction and cardiac dysfunction," *Europace*, vol. 11, pp. 514–520, 2009.

143. L. De Ambroggi and A. D. Corlan, "Clinical use of body surface potential mapping in cardiac arrhythmias," *Anadolu. Kardiyol. Der.*, vol. 7 Suppl 1, pp. 8–10, 2007.

144. "EU Heart Project." http://www.euheart.eu/.

145. Polhemus, *3SPACE / FASTRAK User's Manual*. Colchester, Vermont, 2005.

146. D. Farina, *Forward and Inverse Problems of Electrocardiography: Clinical Investigations*. PhD thesis, 2008.

147. A. L. Hodgkin and A. F. Huxley, "A quantitative description of membrane current and its application to conduction and excitation in nerve," *J. Physiol.*, vol. 177, pp. 500–544, 1952.

148. L. Livshitz, K. Decker, G. Faber, T. O'Hara, J. Silva, and Y. Rudy, "Comments on "A model for human ventricular tissue"," *Am. J. Physiol-Heart C.*, vol. 288, pp. H453; author reply H453–4, 2005.

149. C. H. Luo and Y. Rudy, "A dynamic model of the cardiac ventricular action potential. I. Simulations of ionic currents and concentraion changes," *Circ. Res.*, vol. 74, pp. 1071–1096, 1994.

150. J. Zeng, K. R. Laurita, D. S. Rosenbaum, and Y. Rudy, "Two components of the delayed rectifier K+ current in ventricular myocytes of the guinea pig type. Theoretical formulation and their role in repolarization," *Circ. Res.*, vol. 77, pp. 140–152, 1995.

151. K. H. W. J. ten Tusscher, D. Noble, P. J. Noble, and A. V. Panfilov, "A Model for Human Ventricular Tissue," *Am. J. Physiol.*, vol. 286, pp. H1573–H1589, 2004.

152. K. H. W. J. ten Tusscher and A. V. Panfilov, "Alternans and spiral breakup in a human ventricular tissue model," *Am. J. Physiol-Heart C.*, vol. 291, pp. H1088–100, 2006.

153. E. Grandi, F. S. Pasqualini, and D. M. Bers, "A novel computational model of the human ventricular action potential and Ca transient," *J. Mol. Cell. Cardiol.*, vol. 48, pp. 112–121, 2010.

154. F. G. Akar, G. X. Yan, C. Antzelevitch, and D. S. Rosenbaum, "Unique topographical distribution of M cells underlies reentrant mechanism of torsade de pointes in the long-QT sydrome," *Circulation*, vol. 105, pp. 1247–1253, 2002.

155. A. V. Glukhov, V. V. Fedorov, Q. Lou, V. K. Ravikumar, P. W. Kalish, R. B. Schuessler, N. Moazami, and I. R. Efimov, "Transmural dispersion of repolarization in failing and nonfailing human ventricle," *Circ. Res.*, vol. 106, pp. 981–991, 2010.

156. K. Harumi, M. J. Burgess, and J. A. Abildskov, "A theoretic model of the T wave," *Circulation*, vol. 34, pp. 657–668, 1966.

157. W. T. Miller and D. B. Geselowitz, "Simulation studies of the electrocardiogram. I. The normal heart," *Circ. Res.*, vol. 43, pp. 301–315, 1986.

158. D. di Bernardo and A. Murray, "Computer model for study of cardiac repolarization," *J. Cardiovasc. Electr.*, vol. 11, pp. 895–899, 2000.

159. J. Xue, Y. Chen, X. Han, and W. Gao, "Electrocardiographic morphology changes with different type of repolarization dispersions," *J. Electrocardiol.*, vol. 43, pp. 553–559, 2010.

160. J. Xue, W. Gao, Y. Chen, and X. Han, "Study of repolarization heterogeneity and electrocardiographic morphology with a modeling approach," *J. Electrocardiol.*, vol. 41, pp. 581–587, 2008.

161. J. Xue, W. Gao, Y. Chen, and X. Han, "Identify drug-induced T wave morphology changes by a cell-to-electrocardiogram model and validation with clinical trial data," *J. Electrocardiol.*, vol. 42, pp. 534–542, 2009.

162. M. Potse, A. R. LeBlanc, and A. Vinet, "Why do we need supercomputers to understand the electrocardiographic T wave?," *Anadolu. Kardiyol. Der.*, vol. 7 Suppl 1, pp. 123–124, 2007.

163. P. Viswanathan, R. M. Shaw, and Y. Rudy, "Effects of IKr and IKs Heterogeneity on Action Potential Duration and its Rate Dependence: A Simulation Study," *Circulation*, vol. 99, pp. 2466–2474, 1999.

164. C. E. Conrath, R. Wilders, R. Coronel, J. M. d. Bakker, P. Taggart, J. R. d. Groot, and T. Opthof, "Intercellular coupling through gap junctions masks M cells in the human heart," *Cardiovasc. Res.*, vol. 62, pp. 407–414, 2004.

165. P. Colli Franzone, L. F. Pavarino, S. Scacchi, and B. Taccardi, "T wave polarity of simulated electrocardiograms: influence of transmural heterogeneity," *Int. J. Bioelectromagnetism*, vol. 11, pp. 11–16, 2009.

166. M. Potse, B. Dubé, J. Richter, A. Vinet, and R. M. Gulrajani, "A Comparison of Monodomain and Bidomain Reaction-Diffusion Models for Action Potential Propagation in the Human Heart," *IEEE Trans. Biomed. Eng.*, vol. 53, pp. 2425–2435, 2006.

167. K. J. Sampson and C. S. Henriquez, "Electrotonic influences on action potential duration dispersion in small hearts: a simulation study," *Am. J. Physiol-Heart C.*, vol. 289, pp. H350–60, 2005.

168. M. Potse, A. Vinet, T. Opthof, and R. Coronel, "Validation of a simple model for the morphology of the T wave in unipolar electrograms," *Am. J. Physiol-Heart C.*, vol. 297, pp. H792–801, 2009.

169. P. Colli Franzone, L. F. Pavarino, S. Scacchi, and B. Taccardi, "Modeling ventricular repolarization: effects of transmural and apex-to-base heterogeneities in action potential durations," *Math. Biosci.*, vol. 214, pp. 140–152, 2008.

170. J. G. Stinstra, B. Hopenfeld, and R. S. MacLeod, "On the passive cardiac conductivity," *Ann. Biomed. Eng.*, vol. 33, pp. 1743–1751, 2005.

171. C. S. Henriquez, "Simulating the electrical behavior of cardiac tissue using the bidomain model," *Crit. Rev. Biomed. Eng.*, vol. 21, pp. 1–77, 1993.

172. B. M. Rocha, F. Kickinger, A. J. Prassl, G. Haase, E. J. Vigmond, R. W. D. Santos, S. Zaglmayr, and G. Plank, "A macro finite-element formulation for cardiac electrophysiology simulations using hybrid unstructured grids," *IEEE Trans. Biomed. Eng.*, vol. 58, pp. 1055–1065, 2011.

173. M. J. Bishop and G. Plank, "Representing cardiac bidomain bath-loading effects by an augmented monodomain approach: application to complex ventricular models," *IEEE Trans. Biomed. Eng.*, vol. 58, pp. 1066–1075, 2011.

174. D. Durrer, R. T. van Dam, G. E. Freud, M. J. Janse, F. L. Meijler, and R. C. Arzbaecher, "Total excitation of the isolated human heart," *Circulation*, vol. 41, pp. 899–912, 1970.

175. A. E. Pollard and R. C. Barr, "The construction of an anatomically based model of the human ventricular conduction system," *IEEE Trans. Biomed. Eng.*, vol. 37, pp. 1173–1185, 1990.

176. A. E. Pollard and R. C. Barr, "Computer simulations of activation in an anatomically based model of the human ventricular conduction system," *IEEE Trans. Biomed. Eng.*, vol. 38, pp. 982–996, 1991.

177. P. Siregar, J. P. Sinteff, N. Julen, and P. L. Beux, "An Interactive 3D Anisotropic Cellular Automata Model of the Heart," *Comput. Biomed. Res.*, vol. 31, pp. 323–347, 1998.

178. O. Berenfeld and J. Jalife, "Purkinje-muscle reentry as a mechanism of polymorphic ventricular arrhythmias in a 3-dimensional model of the ventricles," *Circ. Res.*, vol. 82, pp. 1063–1077, 1998.

179. K. Simelius, J. Nenonen, and B. M. Horacek, "Modeling Cardiac Ventricular Activation," *Int. J. Bioelectromagnetism*, vol. 3, 2001.

180. E. J. Vigmond and C. Clements, "Construction of a Computer Model to Investigate Sawtooth Effects in the Purkinje System," *IEEE Trans. Biomed. Eng.*, vol. 54, pp. 389–399, 2007.

181. K. H. W. J. T. Tusscher and A. V. Panfilov, "Modelling of the ventricular conduction system," *Progr. Biophys. Mol. Biol.*, vol. 96, pp. 152–170, 2008.

182. S. Abboud, O. Berenfeld, and D. Sadeh, "Simulation of high-resolution QRS complex using a ventricular model with a fractal conduction system. Effects of ischemia on high-frequency QRS potentials," *Circ. Res.*, vol. 68, pp. 1751–1760, 1991.

183. M. Lorange and R. M. Gulrajani, "A computer heart model incorporating anisotropic propagation," *J. Electrocardiol.*, vol. 26, pp. 245–261, 1993.

184. O. Berenfeld and S. Abboud, "Simulation of cardiac activity and the ECG using a heart model with a reaction-diffusion action potential," *Med. Eng. Phys.*, vol. 18, pp. 615–625, 1996.

185. C. D. Werner, F. B. Sachse, and O. Dössel, "Electrical Excitation Propagation in the Human Heart," *Int. J. Bioelectromagnetism*, vol. 2, 2000.

186. T. Ijiri, T. Ashihara, T. Yamaguchi, K. Takayama, T. Igarashi, T. Shimada, T. Namba, R. Haraguchi, and K. Nakazawa, "A procedural method for modeling the Purkinje fibers of the heart," *J. Physiol. Sci.*, vol. 58, pp. 481–486, 2008.

187. A. Lindenmayer, "Mathematical models for cellular interactions in development. I&II," *J. Theor. Biol.*, vol. 18, pp. 280–315, 1968.

188. V. Zimmerman, R. Sebastian, B. H. Bijnens, and A. F. Frangi, "Modeling the Purkinje conduction system with a non deterministic rule based iterative method," in *Proc. CinC*, vol. 36, pp. 461–464, 2009.

189. D. Romero, V. Zimmerman, R. Sebastian, and A. F. Frangi, "Flexible modeling for anatomically-based cardiac conduction system construction," *Conf. Proc. IEEE Eng. Med. Biol. Soc.*, vol. 2010, pp. 779–782, 2010.

190. D. Romero, R. Sebastian, B. H. Bijnens, V. Zimmerman, P. M. Boyle, E. J. Vigmond, and A. F. Frangi, "Effects of the purkinje system and cardiac geometry on biventricular pacing: a model study," *Ann. Biomed. Eng.*, vol. 38, pp. 1388–1398, 2010.

191. O. Camara, A. Pashaei, R. Sebastian, and A. F. Frangi, "Personalization of fast conduction Purkinje system in eikonal-based electrophysiological models with optical mapping data," in *LNCS*, vol. 6364, pp. 281–290, 2010.

192. R. Bordas, V. Grau, R. B. Burton, P. Hales, J. E. Schneider, D. Gavaghan, P. Kohl, and B. Rodriguez, "Integrated approach for the study of anatomical variability in the cardiac Purkinje system: from high resolution MRI to electrophysiology simulation," *Conf. Proc. IEEE Eng. Med. Biol. Soc.*, vol. 2010, pp. 6793–6796, 2010.

193. H. E. Cetingül, G. Plank, N. A. Trayanova, and R. Vidal, "Estimation of Local Orientations in Fibrous Structures with Applications to the Purkinje System," *IEEE Trans. Biomed. Eng.*, 2011.

194. P. M. Boyle, M. Deo, G. Plank, and E. J. Vigmond, "Purkinje-mediated effects in the response of quiescent ventricles to defibrillation shocks," *Ann. Biomed. Eng.*, vol. 38, pp. 456–468, 2010.

195. M. Deo, P. Boyle, G. Plank, and E. J. Vigmond, "Role of Purkinje system in cardiac arrhythmias," *Conf. Proc. IEEE Eng. Med. Biol. Soc.*, vol. 2008, pp. 149–152, 2008.

196. M. Deo, P. Boyle, G. Plank, and E. J. Vigmond, "Arrhythmogenic mechanisms of the Purkinje system during electric shocks: a modeling study," *Heart Rhythm*, vol. 6, pp. 1782–1789, 2009.

197. D. Noble, "A modification of the Hodgkin-Huxley equation applicable to Purkinje fibre action and pacemaker potentials," *J. Physiol.*, vol. 160, pp. 317–352, 1962.

198. R. E. McAllister, D. Noble, and R. W. Tsien, "Reconstruction of the electrical activity of cardiac Purkinje fibres," *Journal of Physiology*, vol. 251, pp. 1–59, 1975.

199. D. DiFrancesco and D. Noble, "A model of cardiac electrical activity incorporating ionic pumps and concentration changes," *Phil. Trans. R. Soc. Lond.*, vol. 307, pp. 353–398, 1985.

200. P. Stewart, O. V. Aslanidi, D. Noble, P. J. Noble, M. R. Boyett, and H. Zhang, "Mathematical models of the electrical action potential of Purkinje fibre cells," *Philos. T. Roy. Soc. A*, vol. 367, pp. 2225–2255, 2009.

201. E. Grandi, S. Vecchietti, S. Severi, E. Giordano, and S. Cavalcanti, "Effects of beta-adrenergic stimulation on the ventricular action potential: a simulation study," in *International Conference Modelling in Medicine and Biology VI, Bologna* (M. Ursino, ed.), pp. 87–94, WIT Press Southampton, 2005.

202. R. Wilders, M. Hoekstra, A. C. G. van Ginneken, and A. O. Verkerk, "Beta-adrenergic modulation of heart rate: contribution of the slow delayed rectifier K+ current (IKs)," in *Proc. CinC*, vol. 37, pp. 629–631, 2010.

203. J. Koivumaeki, T. Korhonen, J. Takalo, M. Weckstroem, and P. Tavi, "It Takes Two to Tango: Regulation of Sarcoplasmic Reticulum Calcium ATPase by CaMK and PKA in a Mouse Cardiac Myocyte," in *Proc. IFMBE*, vol. 22, pp. 2699–2702, 2009.

204. A. G. Gilman, M. I. Simon, H. R. Bourne, B. A. Harris, R. Long, E. M. Ross, J. T. Stull, R. Taussig, A. P. Arkin, M. H. Cobb, J. G. Cyster, P. N. Devreotes, J. E. Ferrell, D. Fruman, M. Gold, A. Weiss, M. J. Berridge, L. C. Cantley, W. A. Catterall, S. R. Coughlin, E. N. Olson, T. F. Smith, J. S. Brugge, D. Botstein, J. E. Dixon, T. Hunter, R. J. Lefkowitz, A. J. Pawson, P. W. Sternberg, H. Varmus, S. Subramaniam, R. S. Sinkovits, J. Li, D. Mock, Y. Ning, B. Saunders, P. C. Sternweis, D. Hilgemann, R. H. Scheuermann, D. DeCamp, R. Hsueh, K. M. Lin, Y. Ni, W. E. Seaman, P. C. Simpson, T. D. O'Connell, T. Roach, S. Choi, P. Eversole-Cire, I. Fraser, M. C. Mumby, Y. Zhao, D. Brekken, H. Shu, T. Meyer, G. Chandy, W. D. Heo, J. Liou, N. O'Rourke, M. Verghese, S. M. Mumby, H. Han, H. A. Brown, J. S. Forrester, P. Ivanova, S. B. Milne, P. J. Casey, T. K. Harden, J. Doyle, M. L. Gray, S. Michnick, M. A. Schmidt, M. Toner, R. Y. Tsien, M. Natarajan, R. Ranganathan, and G. R. Sambrano, "Overview of the Alliance for Cellular Signaling," *Nature*, vol. 420, pp. 703–706, 2002.

205. J. J. Saucerman, L. L. Brunton, A. P. Michailova, and A. D. McCulloch, "Modeling beta-adrenergic control of cardiac myocyte contractility in silico," *J. Biol. Chem.*, vol. 278, pp. 47997–48003, 2003.

206. J. L. Puglisi and D. M. Bers, "LabHEART: An Interactive Computer Model of Rabbit Ventricular Myocyte Ion Channels and Ca transport," *Am. J. Physiol.*, vol. 281, pp. 2049–2060, 2001.

207. D. M. Bers, *Excitation-contraction coupling and cardiac contractile force*, vol. 2 of *Developments in cardiovascular medicine*. Boston: Kluwer / Academic Publishers, 2001.

208. M. C. Jafri, J. J. Rice, and R. L. Winslow, "Cardiac Ca2+ Dynamics: The roles of ryanodine receptor adaption and sarcoplasmic reticulum load," *Biophys. J.*, vol. 74, pp. 1149–1168, 1998.

209. S. V. Pandit, R. B. Clark, W. R. Giles, and S. S. Demir, "A mathematical model of action potential heterogeneity in adult rat left ventricular myocytes," *Biophys. J.*, vol. 81, pp. 3029–3051, 2001.

210. J. J. Saucerman and A. D. McCulloch, "Mechanistic systems models of cell signaling networks: a case study of myocyte adrenergic regulation," *Prog. Biophys. Mol. Bio.*, vol. 85, pp. 261–278, 2004.

211. Y. Himeno, N. Sarai, S. Matsuoka, and A. Noma, "Ionic mechanisms underlying the positive chronotropy induced by beta1-adrenergic stimulation in guinea pig sinoatrial node cells: a simulation study," *J. Physiol. Sci.*, vol. 58, pp. 53–65, 2008.

212. M. Kuzumoto, A. Takeuchi, H. Nakai, C. Oka, A. Noma, and S. Matsuoka, "Simulation analysis of intracellular Na+ and Cl- homeostasis during beta 1-adrenergic stimulation of cardiac myocyte," *Prog. Biophys. Mol. Bio.*, vol. 96, pp. 171–186, 2008.

213. A. R. Soltis and J. J. Saucerman, "Synergy between CaMKII substrates and beta-adrenergic signaling in regulation of cardiac myocyte Ca(2+) handling," *Biophys. J.*, vol. 99, pp. 2038–2047, 2010.

214. D. M. Bers, "Calcium cycling and signaling in cardiac myocytes," *Annu. Rev. Physiol.*, vol. 70, pp. 23–49, 2008.

215. S. Sicouri, D. Antzelevitch, C. Heilmann, and C. Antzelevitch, "Effects of sodium channel block with mexiletine to reverse action potential prolongation in in vitro models of the long term QT syndrome," *J. Cardiovasc. Electr.*, vol. 8, pp. 1280–1290, 1997.

216. G. X. Yan and C. Antzelevitch, "Cellular basis for the normal T wave and the electrocardiographic manifestations of the long-QT syndrome," *Circulation*, vol. 98, pp. 1928–1936, 1998.

217. W. Shimizu and C. Antzelevitch, "Cellular basis for the ECG features of the LQT1 form of the long-QT syndrome: Effects of β-adrenergic agonists and antagonists and sodium channel blockers on transmural dispersion of repolarization and torsade de pointes," *Circulation*, vol. 98, pp. 2314–2322, 1998.

218. W. Shimizu and C. Antzelevitch, "Cellular basis for long QT, transmural dispersion of repolarization, and torsade de pointes in the long QT syndrome," *J. Electrocardiol.*, vol. 32 Suppl, pp. 177–184, 1999.

219. W. Shimizu, B. McMahon, and C. Antzelevitch, "Sodium pentobarbital reduces transmural dispersion of repolarization and prevents torsades de Pointes in models of acquired and congenital long QT syndrome," *J. Cardiovasc. Electr.*, vol. 10, pp. 154–164, 1999.

220. W. Shimizu and C. Antzelevitch, "Cellular and ionic basis for T-wave alternans under long-QT conditions," *Circulation*, vol. 99, pp. 1499–1507, 1999.

221. W. Shimizu and C. Antzelevitch, "Differential effects of beta-adrenergic agonists and antagonists in LQT1, LQT2 and LQT3 models of the long QT syndrome," *J. Am. Coll. Cardiol.*, vol. 35, pp. 778–786, 2000.

222. W. Shimizu and C. Antzelevitch, "Effects of a K(+) channel opener to reduce transmural dispersion of repolarization and prevent torsade de pointes in LQT1, LQT2, and LQT3 models of the long-QT syndrome," *Circulation*, vol. 102, pp. 706–712, 2000.

223. W. Shimizu, T. Aiba, and C. Antzelevitch, "Specific therapy based on the genotype and cellular mechanism in inherited cardiac arrhythmias. Long QT syndrome and Brugada syndrome," *Curr. Pharm. Design*, vol. 11, pp. 1561–1572, 2005.

224. X. H. Wehrens, H. Abriel, C. Cabo, J. Benhorin, and R. S. Kass, "Arrhythmogenic mechanism of an LQT-3 mutation of the human heart Na(+) channel alpha-subunit: A computational analysis," *Circulation*, vol. 102, pp. 584–590, 2000.

225. K. J. Sampson, V. Iyer, A. R. Marks, and R. S. Kass, "A computational model of Purkinje fibre single cell electrophysiology: implications for the long QT syndrome," *The Journal of Physiology*, vol. 588, pp. 2643–2655, 2010.

226. M. Potse, G. Baroudi, P. A. Lanfranchi, and A. Vinet, "Generation of the T wave in the electrocardiogram: lessons to be learned from long-QT syndromes," in *Can. J. Cardiol.*, vol. 23 Suppl., p. C:238C, 2007.

227. G. Seemann, D. L. Weiß, F. B. Sachse, and O. Dössel, "Simulation of the Long-QT Syndrome in a Model of Human Myocardium," in *Proc. CinC*, vol. 30, pp. 287–290, 2003.

228. R. H. Clayton, A. Bailey, V. N. Biktashev, and A. V. Holden, "Re-entrant cardiac arrhythmias in computational models of long QT myocardium," *J. Theor. Biol.*, vol. 208, p. 215, 2001.

229. R. C. Ahrens-Nicklas, C. E. Clancy, and D. J. Christini, "Re-evaluating the efficacy of beta-adrenergic agonists and antagonists in long QT-3 syndrome through computational modelling," *Cardiovasc. Res.*, vol. 82, pp. 439–447, 2009.

230. E. M. Cherry, H. S. Greenside, and C. S. Henriquez, "Efficient simulation of three-dimensional anisotropic cardiac tissue using an adaptive mesh refinement method," *Chaos*, vol. 13, pp. 853–865, 2003.

231. V. G. Fast and A. G. Kleber, "Role of wavefront curvature in propagation of cardiac impulse," *Cardiovasc. Res.*, vol. 33, pp. 258–271, 1997.

232. J. M. Peyrat, M. Sermesant, X. Pennec, H. Delingette, C. Xu, E. McVeigh, and N. Ayache, "Towards a statistical atlas of cardiac fiber structure," *Proc. MICCAI*, vol. 9, pp. 297–304, 2006.

233. J. M. Peyrat, M. Sermesant, X. Pennec, H. Delingette, C. Xu, E. R. McVeigh, and N. Ayache, "A computational framework for the statistical analysis of cardiac diffusion tensors: application to a small database of canine hearts," *IEEE Trans. Med. Imaging*, vol. 26, pp. 1500–1514, 2007.

234. D. U. J. Keller, G. Seemann, D. L. Weiss, and O. Dössel, "Detailed anatomical modeling of human ventricles based on diffusion tensor MRI," in *Proc. BMT*, vol. 50/1, 2006.

235. D. L. Weiss, D. U. J. Keller, G. Seemann, and O. Dössel, "The influence of fibre orientation, extracted from different segments of the human left ventricle, on the activation and repolarization sequence: a simulation study," *Europace*, vol. 9, p. 96, 2007.

236. D. U. J. Keller, D. L. Weiss, O. Dössel, and G. Seemann, "Transferring ventricular myocyte orientation to individual patient data based on diffusion tensor MRI measurements," in *Proc. CURAC*, pp. 255–258, 2007.

237. N. Toussaint, M. Sermesant, C. T. Stoeck, S. Kozerke, and P. G. Batchelor, "In vivo human 3D cardiac fibre architecture: reconstruction using curvilinear interpolation of diffusion tensor images," *Proc. MICCAI*, vol. 13, pp. 418–425, 2010.

238. M. L. Buist and A. J. Pullan, "The effect of torso impedance on epicardial and body surface potentials: a modeling study," *IEEE Trans. Biomed. Eng.*, vol. 50, pp. 816–824, 2003.

239. O. Skipa, *Linear inverse problem of electrocardiography: epicardial potentials and transmembrane voltages*. PhD thesis, 2004.

240. C. P. Bradley, A. J. Pullan, and P. J. Hunter, "Effects of material properties and geometry on electro-cardiographic forward simulations," *Ann. Biomed. Eng.*, vol. 28, pp. 721–741, 2000.

241. K. R. Foster and H. P. Schwan, "Dielectric properties of tissues and biological materials: A critical review," *Crit. Rev. Biomed. Eng.*, vol. 17, pp. 25–104, 1989.

242. S. Gabriel, R. W. Lau, and C. Gabriel, "The dielectric properties of biological tissues: II. Measurements in the frequency range 10 Hz to 20 GHz," *Phys. Med. Biol.*, vol. 41, pp. 2251–22269, 1996.

243. R. N. Klepfer, C. R. Johnson, and R. S. MacLeod, "The Effects of Inhomogeneities and Anisotropies on Electrocardiographic Fields: A 3-D Finite-Element Study," *IEEE Trans. Biomed. Eng.*, vol. 44, pp. 706–719, 1997.

244. T. J. C. Faes, H. A. van der Meij, J. C. de Munck, and R. M. Heethaar, "The electric resistivity of human tissues (100 Hz-10 MHz) a meta-analysis of review studies," *Physiological Measurement*, vol. 20, p. 1, 1999.

245. S. Gabriel, R. W. Lau, and C. Gabriel, "The dielectric properties of biological tissues: III. Parametric models for the dielectric spectrum of tissues," *Phys. Med. Biol.*, vol. 41, pp. 2271–2293, 1996.

246. F. A. Duck, *Physical properties of tissue: A comprehensive reference book*. Academic Press, London San Diego, 1990.

247. H. C. Burger and R. van Dongen, "Specific electric resistance of body tissues," *Phys. Med. Biol.*, vol. 5, pp. 431–447, 1961.

248. L. A. Geddes and L. E. Baker, "The specific resistance of biological material-A compendium of data for the biomedical engineer and physiologist," *Med. Biol. Eng.*, vol. 5, pp. 271–293, 1967.

249. E. Zheng, S. Shao, and J. G. Webster, "Impedance of skeletal muscle from 1 Hz to 1 MHz," *IEEE Trans. Biomed. Eng.*, vol. BME-31, pp. 477–481, 1984.

250. A. van Oosterom, R. W. de Boer, and R. T. van Dam, "Intramural resistivity of cardiac tissue," *Med. Biol. Eng. Comput.*, vol. 17, pp. 337–343, 1979.

251. Y. Rudy, R. Plonsey, and J. Liebman, "The effects of variations in conductivity and geometrical parameters on the electrocardiogram, using an eccentric spheres model," *Circ. Res.*, vol. 44, pp. 104–111, 1979.

252. C. R. Johnson, R. S. MacLeod, and P. R. Ershler, "A computer model for the study of electrical current flow in the human thorax," *Comput. Biol. Med.*, vol. 22, pp. 305–323, 1992.

253. F. B. Sachse, M. Wolf, C. D. Werner, and K. Meyer-Waarden, "Extension of anatomical models of the human body: Three-dimensional interpolation of muscle fiber orientation based on restrictions," *Journal of Computing and Information Technology*, vol. 6, pp. 95–101, 1998.

254. R. S. MacLeod, R. L. Lux, and B. Taccardi, "A possible mechanism for electrocardiographically silent changes in cardiac repolarization," *J. Electrocardiol.*, vol. 30, pp. 114–121, 1998.

255. T. Feldman, R. W. Childers, K. M. Borow, R. M. Lang, and A. Neumann, "Change in ventricular cavity size: differential effects on QRS and T wave amplitude," *Circulation*, vol. 72, pp. 495–501, 1985.

256. R. M. Gulrajani and G. E. Mailloux, "A simulation study of the effects of torso inhomogeneities on electrocardiographic potentials, using realistic heart and torso models," *Circ. Res.*, vol. 52, pp. 45–56, 1983.

257. D. A. Brody, "A theoretical analysis of intracavitary blood mass influence on the heart-lead relationship," *Circ. Res.*, vol. 4, pp. 731–738, 1956.

258. L. Xia, M. Huo, Q. Wei, F. Liu, and S. Crozier, "Analysis of cardiac ventricular wall motion based on a three-dimensional electromechanical biventricular model," *Phys. Med. Biol.*, vol. 50, pp. 1901–1917, 2005.

259. Q. Wei, F. Liu, B. Appleton, L. Xia, N. Liu, S. Wilson, R. Riley, W. Strugnel, R. Slaughter, R. Denman, and S. Crozier, "Effect of cardiac motion on body surface electrocardiographic potentials: an MRI-based simulation study," *Phys. Med. Biol.*, vol. 51, pp. 3405–3418, 2006.

260. N. P. Smith, M. L. Buist, and A. J. Pullan, "Altered T wave dynamics in a contracting cardiac model," *J. Cardiovasc. Electrophysiol.*, vol. 14, pp. S203–9, 2003.

261. E. Drouin, F. Charpentier, C. Gauthier, K. Laurent, and H. L. Marec, "Electrophysiological characteristics of cells spanning the left ventricular wall of human heart: Evidence for the presence of M Cells," *J. Am. Coll. Cardiol.*, vol. 26, pp. 185–192, 1995.

262. G. X. Yan, W. Shimizu, and C. Antzelevitch, "Characteristics and distribution of M cells in arterially perfused canine left ventricular wedge preparations," *Circulation*, vol. 98, pp. 1921–1927, 1998.

263. N. Szentadrassy, T. Banyasz, T. Biro, G. Szabo, B. I. Toth, J. Magyar, J. Lazar, A. Varro, L. Kovacs, and P. P. Nanasi, "Apico-basal inhomogeneity in distribution of ion channels in canine and human ventricular myocardium," *Cardiovasc. Res.*, vol. 65, pp. 851–860, 2005.

264. A. Khawaja, S. Sanyal, and O. Dössel, "A wavelet-based technique for baseline wander correction in ECG and multi-channel ECG," in *Proc. IFMBE*, vol. 9, pp. 291–292, 2005.

265. A. Khawaja, *Automatic ECG analysis using principal component analysis and wavelet transformation*, vol. 3 of *Karlsruhe Transactions on Biomedical Engineering*. Karlsruhe: Universitätsverlag Karlsruhe, 2006.

266. "CGAL, Computational Geometry Algorithms Library." http://www.cgal.org.

267. S. Rush and H. Larsen, "A practical algorithm for solving dynamic membrane equations," *IEEE Trans. Biomed. Eng.*, vol. 25, pp. 389–392, 1978.

268. D. W. Liu and C. Antzelevitch, "Characteristics of the delayed rectifier current (IKr and IKs) in canine ventricular epicardial, midmyocardial, and endocardial myocytes. A weaker IKs contributes to the longer action potential of the M cell," *Circ. Res.*, vol. 76, pp. 351–365, 1995.

269. S. Sicouri and C. Antzelevitch, "Distribution of M cells in the canine ventricle," *J. Cardiovasc. Electr.*, vol. 5, pp. 824–837, 1994.

270. S. Sicouri and C. Antzelevitch, "A subpopulation of cells with unique electrophysiological properties in the deep subepicardium of the canine ventricle: The M Cell," *Circ. Res.*, vol. 68, pp. 1729–1741, 1991.

271. S. Sicouri, A. Glass, M. Ferreiro, and C. Antzelevitch, "Transseptal dispersion of repolarization and its role in the development of Torsade de Pointes arrhythmias," *J. Cardiovasc. Electr.*, vol. 21, pp. 441–447, 2010.

272. F. G. Akar and D. S. Rosenbaum, "Transmural electrophysiological heterogeneities underlying arrhythmogenesis in heart failure," *Circ. Res.*, vol. 93, pp. 638–645, 2003.

273. E. P. Anyukhovsky, E. A. Sosunov, and M. R. Rosen, "Regional differences in electrophysiological properties of epicardium, midmyocardium, and endocardium," *Circulation*, vol. 94, pp. 1981–1988, 1996.

274. T. Stankovicova, M. Szilard, I. Scheerder, and K. Sipido, "M Cells and transmural heterogeneity of action potential configuration in myocytes from the left ventricular wall of the pig heart," *Cardiovasc. Res.*, vol. 45, pp. 952–960, 2000.

275. S. Sicouri, M. Quist, and C. Antzelevitch, "Evidence for the presence of M Cells in the guinea pig ventricle," *J. Cardiovasc. Electr.*, vol. 7, pp. 503–511, 1996.

276. M. A. McIntosh, S. M. Cobbe, and G. L. Smith, "Heterogeneous changes in action potential and intracellular Ca2+ in left ventricular myocyte sub-types from rabbits with heart failure," *Cardiovasc. Res.*, vol. 45, pp. 397–409, 2000.

277. G. Szabo, N. Szentandrassy, T. Biro, B. I. Toth, G. Czifra, J. Magyar, T. Banyasz, A. Varro, L. Kovacs, and P. P. Nanasi, "Asymmetrical distribution of ion channels in canine and human left-ventricular wall: epicardium versus midmyocardium," *Pflugers. Arch.*, vol. 450, pp. 307–316, 2005.

278. S. Sekiya, S. Ichikawa, T. Tsutsumi, and K. Harumi, "Distribution of action potential durations in the canine left ventricle," *Jpn. Heart J.*, vol. 25, pp. 181–194, 1984.

279. M. N. Obreztchikova, K. W. Patberg, A. N. Plotnikov, N. Ozgen, I. N. Shlapakova, A. V. Rybin, E. A. Sosunov, P. J. Danilo, E. P. Anyukhovsky, R. B. Robinson, and M. R. Rosen, "I(Kr) contributes to the altered ventricular repolarization that determines long-term cardiac memory," *Cardiovasc. Res.*, vol. 71, pp. 88–96, 2006.

280. M. J. Burgess, L. S. Green, K. Millar, R. Wyatt, and J. A. Abildskov, "The sequence of normal ventricular recovery," *Am. Heart J.*, vol. 84, pp. 660–669, 1972.

281. C. Patel, J. F. Burke, H. Patel, P. Gupta, P. R. Kowey, C. Antzelevitch, and G. X. Yan, "Is there a significant transmural gradient in repolarization time in the intact heart? Cellular basis of the T wave: a century of controversy," *Circ. Arrhythm. Electrophysiol.*, vol. 2, pp. 80–88, 2009.

282. T. Watanabe, P. M. Rautaharju, and T. F. McDonald, "Ventricular action potentials, ventricular extracellular potentials, and the ECG of guinea pig," *Circ. Res.*, vol. 57, pp. 362–373, 1985.

283. S. M. Bryant, X. Wan, S. J. Shipsey, and G. Hart, "Regional differences in the delayed rectifier current (IKr and IKs) contribute to the differences in action potential duration in basal left ventricular myocytes in guinea-pig," *Cardiovasc. Res.*, vol. 40, pp. 322–331, 1998.

284. M. V. Brahmajothi, M. J. Morales, K. A. Reimer, and H. C. Strauss, "Regional localization of ERG, the channel protein responsible for the rapid current of the delayed rectifier, K^+ current in the ferret heart," *Circ. Res.*, vol. 81, pp. 128–135, 1997.

285. M. J. Janse, E. A. Sosunov, R. Coronel, T. Opthof, E. P. Anyukhovsky, J. M. T. d. Bakker, A. N. Plotnikov, I. N. Shlapakova, P. Danilo, J. G. P. Tijssen, and M. R. Rosen, "Repolarization Gradients in the Canine Left Ventricle Before and After Induction of Short-Term Cardiac Memory," *Circulation*, vol. 112, pp. 1711–1718, 2005.

286. A. Bauer, R. Becker, C. Karle, K. D. Schreiner, J. C. Senges, F. Voss, P. Kraft, W. Kuebler, and W. Schoels, "Effects of the I(Kr)-blocking agent dofetilide and of the I(Ks)-blocking agent chromanol 293b on regional disparity of left ventricular repolarization in the intact canine heart," *J. Cardiovasc. Pharm.*, vol. 39, pp. 460–467, 2002.

287. J. Cheng, K. Kamiya, W. Liu, Y. Tsuji, J. Toyama, and I. Kodama, "Heterogeneous distribution of the two components of delayed rectifier K^+ current: a potential mechanism of the proarrhythmic effects of methanesulfonanilide class III agents," *Cardiovasc. Res.*, vol. 43, pp. 135–147, 1999.

288. T. Watanabe, L. Delbridge, J. O. Bustanmante, and T. F. McDonald, "Heterogeneity of the action potential in isolated rat ventricular myocytes and tissue," *Circ. Res.*, vol. 52, pp. 280–290, 1983.

289. M. A. Vos, S. H. de Groot, S. C. Verduyn, J. van der Zande, H. D. Leunissen, J. P. Cleutjens, M. van Bilsen, M. J. Daemen, J. J. Schreuder, M. A. Allessie, and H. J. Wellens, "Enhanced susceptibility for acquired torsade de pointes arrhythmias in the dog with chronic, complete AV block is related to cardiac hypertrophy and electrical remodeling," *Circulation*, vol. 98, pp. 1125–1135, 1998.

290. F. H. Samie, O. Berenfeld, J. Anumonwo, S. F. Mironov, S. Udassi, J. Beaumont, S. Taffet, A. M. Pertsov, and J. Jalife, "Rectification of the background potassium current: a determinant of rotor dynamics in ventricular fibrillation," *Circ. Res.*, vol. 89, pp. 1216–1223, 2001.

291. V. S. Chauhan, E. Downar, K. Nanthakumar, J. D. Parker, H. J. Ross, W. Chan, and P. Picton, "Increased ventricular repolarization heterogeneity in patients with ventricular arrhythmia vulnerability and cardiomyopathy: a human in vivo study," *Am. J. Physiol-Heart C.*, vol. 290, pp. H79–86, 2006.

292. S. Yuan, O. Kongstad, E. Hertervig, M. Holm, E. Grins, and B. Olsson, "Global repolarization sequence of the ventricular endocardium: monophasic action potential mapping in swine and humans," *PACE*, vol. 24, pp. 1479–1488, 2001.

293. M. R. Franz, K. Bargheer, W. Rafflenbeul, A. Haverich, and P. R. Lichtlen, "Monophasic action potential mapping in human subjects with normal electrocardiograms: direct evidence for the genesis of the T wave," *Circulation*, vol. 75, pp. 379–386, 1987.

294. G. Autenrieth, B. Surawicz, and C. S. Kuo, "Sequence of repolarization on the ventricular surface in the dog," *Am. Heart J.*, vol. 89, pp. 463–469, 1975.

295. J. C. Cowan, C. J. Hilton, C. J. Griffiths, S. Tansuphaswadikul, J. P. Bourke, A. Murray, and R. W. Campbell, "Sequence of epicardial repolarisation and configuration of the T wave," *Brit. Heart J.*, vol. 60, pp. 424–433, 1988.

296. S. M. MacDonnell, G. Garcia-Rivas, J. A. Scherman, H. Kubo, X. Chen, H. Valdivia, and S. R. Houser, "Adrenergic regulation of cardiac contractility does not involve phosphorylation of the cardiac ryanodine receptor at serine 2808," *Circ. Res.*, vol. 102, pp. e65–72, 2008.

297. Y. Li, E. G. Kranias, G. A. Mignery, and D. M. Bers, "Protein Kinase A phosphorylation of the Ryanodine receptor does not affect calcium sparks in mouse ventricular myocytes," *Circ. Res.*, pp. 309–316, 2002.

298. S. Lehnart and A. R. Marks, "Regulation of ryanodine receptors in the heart," *Circ. Res.*, vol. 101, pp. 746–749, 2007.

299. S. Despa, A. L. Tucker, and D. M. Bers, "Phospholemman-mediated activation of Na/K-ATPase limits [Na]i and inotropic state during beta-adrenergic stimulation in mouse ventricular myocytes," *Circulation*, vol. 117, pp. 1849–1855, 2008.

300. A. Hindmarsh, B. Brown, K. Grant, S. Lee, R. Serban, D. Shumaker, and Woodward, "Suite of nonlinear and differential/algebraic equation solvers," *ACM T. Math. Software*, pp. 363–396, 2005.

301. D. L. Weiß, G. Seemann, and O. Dössel, "Conditions for Equal Polarity of R and T Wave in Heterogeneous Human Ventricular Tissue," in *Proc. BMT*, vol. 49-2/1, pp. 364–365, 2004.

302. P. Viswanathan and Y. Rudy, "Cellular Arrhythmogenic Effects of Congenital and Acquired Long-QT Syndrome in the Heterogeneous Myocardium," *Circulation*, vol. 101, pp. 1192–1198, 2000.

303. M. Glas, "Segmentation des Visible Man Datensatzes," Master's thesis, 1996.

304. "The message passing interface (mpi) standard." http://www-unix.mcs.anl.gov/mpi.

305. Y. Kurata, I. Hisatome, H. Matsuda, and T. Shibamoto, "Dynamical mechanisms of pacemaker generation in IK1-downregulated human ventricular myocytes: Insights from bifurcation analyses of a mathematical model," *Biophys. J.*, vol. 89, pp. 2865–2887, 2005.

306. G. Seemann, F. B. Sachse, D. L. Weiß, and O. Dössel, "Quantitative reconstruction of cardiac electromechanics in human myocardium: Regional heterogeneity," *J. Cardiovasc. Electrophysiol.*, vol. 14, pp. S219–S228, 2003.

307. C. E. Clancy and Y. Rudy, "Na(+) channel mutation that causes both Brugada and long-QT syndrome phenotypes: a simulation study of mechanism," *Circulation*, vol. 105, pp. 1208–1213, 2002.

308. P. J. Dimbylow, "Induced current densities from low-frequency magnetic fields in a 2 mm resolution, anatomically realistic model of the body," *Phys. Med. Biol.*, vol. 43, pp. 221–230, 1998.

309. C. D. Werner, F. B. Sachse, and O. Dössel, "Applications of the Visible Man Dataset in Electrocardiology: Simulation of the Electrical Excitation Propagation," in *Proc. Second Users Conference of the National Library of Medicine's Visible Human Project*, pp. 69–79, 1998.

310. S. Barros, N. S. Couto, F. B. Sachse, C. D. Werner, and G. Seemann, "Segmentation and tissue-classification of the visible female dataset - thoracic muscles, bones and blood vessels," in *Proc. BMT*, vol. 46-1, pp. 510–511, 2001.

311. N. Couto, S. Barros, F. B. Sachse, C. D. Werner, and G. Seemann, "Segmentation and Tissue-Classification of the Visible Female Dataset - non Cardiac Thoracic Organs," in *Proc. BMT*, vol. 46, pp. 512–513, 2001.

312. O. Ecabert, J. Peters, H. Schramm, C. Lorenz, J. von Berg, M. J. Walker, M. Vembar, M. E. Olszewski, K. Subramanyan, G. Lavi, and J. Weese, "Automatic model-based segmentation of the heart in CT images," *IEEE Trans. Med. Imaging*, vol. 27, pp. 1189–1201, 2008.

313. J. Weese, J. Peters, I. Waechter, R. Kneser, H. Lehmann, O. Ecabert, H. Barschdorf, F. M. Weber, O. Doessel, and C. Lorenz, "The Generation of Patient-Specific Heart Models for Diagnosis and Interventions," in *LNCS*, vol. 6364, pp. 25–35, 2010.

314. M. W. Krueger, V. Schmidt, C. Tobón, F. M. Weber, C. Lorenz, D. U. J. Keller, H. Barschdorf, M. Burdumy, P. Neher, G. Plank, K. Rhode, G. Seemann, D. Sanchez-Quintana, J. Saiz, R. Razavi,

and O. Dössel, "Modeling Atrial Fiber Orientation in Patient-Specific Geometries: A Semi-Automatic Rule-Based Approach," in *Proc. FIMH*, 2011.

315. M. W. Krueger, S. Severi, K. Rhode, S. Genovesi, F. W. Weber, A. Vincenti, P. Fabbrini, G. Seemann, R. Razavi, and O. Dössel, "Alterations of Atrial Electrophysiology related to Hemodialysis Session: Insights from a Multi-Scale Computer Model," *J. Electrocardiol.*, vol. in press, 2011.

316. M. W. Krueger, K. Rhode, F. M. Weber, D. U. J. Keller, D. Caulfield, G. Seemann, B. R. Knowles, R. Razavi, and O. Dössel, "Patient-Specific Volumetric Atrial Models with Electrophysiological Components: A Comparison of Simulations and Measurements," in *Proc. BMT*, vol. Supplement 2010, 2010.

317. P. Zerfass, F. B. Sachse, C. D. Werner, and O. Dössel, "Deformation of surface nets for interactive segmentation of tomographic data," in *Proc. BMT*, vol. 45-1, pp. 483–484, 2000.

318. L. Thomas, F. B. Sachse, C. D. Werner, and O. Dössel, "Topologisch veränderbare dreidimensionale Aktive Konturen in der medizinischen Bildverarbeitung," in *Proc. BMT*, vol. 45-1, pp. 509–510, 2000.

319. C. D. Werner, F. B. Sachse, K. Mühlmann, and O. Dössel, "Modellbasierte Segmentation klinischer MR-Aufnahmen," in *Bildverarbeitung für die Medizin 1998*, (Berlin Heidelberg New York), pp. 274–278, Springer, 1998.

320. N. H. Busch, F. B. Sachse, C. D. Werner, and O. Dössel, "Segmentation klinischer vierdimensionaler magnetresonanztomographischer Aufnahmen mittels Aktiver Kontur Modelle und haptischer Interaktion," in *Proc. BMT*, vol. 43-1, pp. 526–529, 1998.

321. F. B. Sachse, *Modelle des menschlichen Körpers zur Berechnung von physikalischen Feldern.* Aachen: Shaker, 1998.

322. S. Y. Wan and W. E. Higgins, "Symmetric Region Growing," *IEEE Trans. Image Process.*, vol. 12, pp. 1007–1015, 2003.

323. J. C. Russ, *The Image Processing Handbook.* Boca Raton; Ann Arbor; London; Tokyo: CRC Press, 1995.

324. P. Taggart, P. M. Sutton, T. Opthof, R. Coronel, R. Trimlett, W. Pugsley, and P. Kallis, "Inhomogeneous transmural conduction during early ischaemia in patients with coronary artery disease," *J. Mol. Cell. Cardiol.*, vol. 32, pp. 621–630, 2000.

325. G. Seemann, F. . Sachse, M. Karl, D. L. Weiss, V. Heuveline, and O. Dössel, "Framework for modular, flexible and efficient solving the cardiac bidomain equation using PETSc," *Progr. Industr. Math.*, vol. 15, pp. 363–369, 2010.

326. J. Biederer, M. Both, J. Graessner, C. Liess, P. Jakob, M. Reuter, and M. Heller, "Lung morphology: fast MR imaging assessment with a volumetric interpolated breath-hold technique: initial experience with patients," *Radiology*, vol. 226, pp. 242–249, 2003.

327. S. Chang, M. D. Cham, S. Hu, and Y. Wang, "3-T navigator parallel-imaging coronary MR angiography: targeted-volume versus whole-heart acquisition," *Am. J. Roentgenol.*, vol. 191, pp. 38–42, 2008.

328. J. Vollmer, R. Mencl, and H. Mueller, "Improved Laplacian Smoothing of Noisy Surface Meshes," *Comput. Graph. Forum*, vol. 3, pp. 131–138, 1999.

329. J. Modersitzki, *Numerical Methods for Image Registration.* New York: Oxford University Press, 2004.

330. M. Ferrant, S. K. Warfield, C. R. G. Guttmann, R. V. Mulkern, F. A. Jolesz, and R. Kikinis, "3D image matching using a finite element based elastic deformation model," in *Proc. MICCAI*, vol. 1679, pp. 202–209, 1999.

331. Y. Xie, M. Xie, and L. Yang, "A New Non-rigid Image Registration Algorithm Using the Finite-Element Method," *Int. Workshop on Education Technology and Computer Science*, vol. 1, pp. 206–210, 2010.

332. Y. Chen and G. Medioni, "Object modeling by registration of multiple range images," in *International Conference on Robotics and Automation*, vol. 3, pp. 2724–2729, 1991.

333. P. J. Besl and H. D. Mckay, "A method for registration of 3-D shapes," *IEEE Trans. Pattern. Anal. Mach. Intell.*, vol. 14, pp. 239–256, 1992.

334. K. Pulli, "Multiview registration for large data sets," in *Proceedings 3-D Digital Imaging and Modeling*, 1999.

335. J. Feldmar, J. Declerck, G. Malandain, N. Ayache, I. Sophia, P. Epidaure, and S. A. Cedex, "Extension of the ICP Algorithm to non rigid Intensity-based Registration of 3D Volumes," in *Comput. Vis. Image Und.*, pp. 84–93, 1996.

336. D. Kim and D. Kim, "A Fast ICP Algorithm for 3-D Human Body Motion Tracking," *IEEE Signal Process. Lett.*, vol. 17, no. 4, pp. 402–405, 2010.

337. S. Rusinkiewicz and M. Levoy, "Efficient Variants of the ICP Algorithm," in *Third International Conference on 3D Digital Imaging and Modeling (3DIM)*, 2001.

338. B. Amberg, S. Romdhani, and T. Vetter, "Optimal step nonrigid ICP algorithms for surface Registration," in *IEEE Comput. Soc. Conf. Comput. Vis. Pattern Recogn.*, pp. 1–8, 2007.

339. O. Jarrousse, T. Fritz, and O. Dössel, "Implicit time integration in a volumetric mass-spring system for modeling myocardial elastomechanics," in *Proc. IFMBE*, vol. 25/4, pp. 876–879, 2009.

340. S. Poelzing, F. G. Akar, E. Baron, and D. S. Rosenbaum, "Heterogeneous connexin43 expression produces electrophysiological heterogeneities across ventricular wall," *Am. J. Physiol-Heart C.*, vol. 286, pp. H2001–9, 2004.

341. S. Poelzing and D. S. Rosenbaum, "Nature, significance, and mechanisms of electrical heterogeneities in ventricle," *Anat. Rec. Part A*, vol. 280, pp. 1010–1017, 2004.

342. M. Courtemanche, R. J. Ramirez, and S. Nattel, "Ionic mechanisms underlying human atrial action potential properties: Insights from a mathematical model," *Am. J. Physiol.*, vol. 275, p. 301, 1998.

343. A. Hansson, M. Holm, P. Blomstrom, R. Johansson, C. Luhrs, J. Brandt, and S. Olsson, "Right atrial free wall conduction velocity and degree of anisotropy in patients with stable sinus rhythm studied during open heart surgery," *Eur. Heart. J.*, vol. 19, pp. 293–300, 1998.

344. D. M. Harrild and C. S. Henriquez, "A Computer Model of Normal Conduction in the Human Atria," *Circ. Res.*, vol. 87, p. 25, 2000.

345. M. W. Krueger, F. M. Weber, G. Seemann, and O. Dössel, "Semi-automatic segmentation of sinus node, Bachmann's Bundle and Terminal Crest for patient specific atrial models," in *Proc. IFMBE*, vol. 25/4, pp. 673–676, 2009.

346. C. F. Kay and H. P. Schwan, "Specific resistance of body tissues," *Circ. Res.*, vol. 4, pp. 664–670, 1956.

347. W. Kaufman and F. D. Johnston, "The electrical conductivity of the tissues near the heart and its bearing on the distribution of the cardiac action currents," *Am. Heart J.*, vol. 26, pp. 42–54, 1943.

348. H. C. Burger and J. B. van Milaan, "Measurements of the specific resistance of the human body to direct current," *Acta Med. Scand.*, vol. 114, pp. 584–607, 1943.

349. H. P. Schwan and C. F. Kay, "The conductivity of living tissues," *Ann. NY Acad. Sci.*, vol. 65, pp. 1007–1013, 1957.

350. S. Rush, J. A. Abildskov, and R. McFee, "Resistivity of body tissues at low frequencies," *Circ. Res.*, vol. 12, pp. 40–50, 1963.

351. B. R. Epstein and K. R. Foster, "Anisotropy in the dielectric properties of skeletal muscle," *Med. Biol. Eng. Comput.*, vol. 21, pp. 51–55, 1983.

352. R. Aaron, M. Huang, and C. A. Shiffman, "Anisotropy of human muscle via non-invasive impedance measurements," *Phys. Med. Biol.*, vol. 42, pp. 1245–1262, 1997.

353. S. Grimnes and O. G. Martinsen, *Bioimpedance and bioelectricity basics.* San Diego, San Francisco, New York: Academic Press, 2008.

354. I. T. Jolliffe, *Principal component analysis.* Berlin: Springer, 2002.

355. Y. Zhang, D. H. Brooks, M. A. Franceschini, and D. A. Boas, "Eigenvector-based spatial filtering for reduction of physiological interference in diffuse optical imaging," *J. Biomed. Opt.*, vol. 10, p. 11014, 2005.

356. R. L. Lux, A. K. Evans, M. J. Burgess, R. F. Wyatt, and J. A. Abildskov, "Redundancy reduction for improved display and analysis of body surface potential maps. I. Spatial compression," *Circ. Res.*, vol. 49, pp. 186–196, 1981.

357. P. Langley, E. J. Bowers, and A. Murray, "Principal Component Analysis as a Tool for Analyzing Beat-to-Beat Changes in ECG Features: Application to ECG-Derived Respiration," *IEEE Trans. Biomed. Eng.*, vol. 57, pp. 821–829, 2010.

358. V. Monasterio, P. Laguna, and J. P. Martinez, "Multilead analysis of T-wave alternans in the ECG using principal component analysis," *IEEE Trans. Biomed. Eng.*, vol. 56, pp. 1880–1890, 2009.

359. S. Olmos, J. García, R. Jané, and P. Laguna, "ECG signal compression plus noise filtering with truncated orthogonal expansions," *Signal Process.*, vol. 79, pp. 97–115, 1999.

360. J. S. Paul, M. R. Reddy, and V. J. Kumar, "A transform domain SVD filter for suppression of muscle noise artefacts in exercise ECG's," *IEEE Trans. Biomed. Eng.*, vol. 47, pp. 654–663, 2000.

361. G. H. Golub and C. Reinsch, "Singular value decomposition and least squares solutions," *Numerische Mathematik*, vol. 14, pp. 403–420, 1970.

362. T. F. Chan, "An improved algorithm for computing the singular value decomposition," *Trans. Math. Softw*, vol. 8, pp. 84–88, 1982.

363. http://www.gnu.org/software/gsl/.

364. J. A. Nelder and R. Mead, "A simplex method for function minimization," *Comput. J.*, vol. 7, pp. 308–313, 1965.

365. A. v. Oosterom and T. F. Oostendorp, "ECGSIM: an interactive tool for studying the genesis of QRST forms," *Heart*, vol. 90, pp. 165–168, 2004.

366. S. T. Morita, D. P. Zipes, H. Morita, and J. Wu, "Analysis of action potentials in the canine ventricular septum: no phenotypic expression of M cells," *Cardiovasc. Res.*, vol. 74, pp. 96–103, 2007.

367. A. Bauer, R. Becker, K. D. Freigang, J. C. Senges, F. Voss, A. Hansen, M. Muller, H. J. Lang, U. Gerlach, A. Busch, P. Kraft, W. Kubler, and W. Schols, "Rate- and site-dependent effects of propafenone, dofetilide, and the new I(Ks)-blocking agent chromanol 293b on individual muscle layers of the intact canine heart," *Circulation*, vol. 100, pp. 2184–2190, 1999.

368. F. Voss, T. Opthof, J. Marker, A. Bauer, H. A. Katus, and R. Becker, "There is no transmural heterogeneity in an index of action potential duration in the canine left ventricle," *Heart Rhythm*, vol. 6, pp. 1028–1034, 2009.

369. A. Rodriguez-Sinovas, J. Cinca, A. Tapias, L. Armadans, M. Tresanchez, and J. Soler-Soler, "Lack of evidence of M-cells in porcine left ventricular myocardium," *Cardiovasc. Res.*, vol. 33, pp. 307–313, 1997.

370. W. Kong, N. Fakhari, O. F. Sharifov, R. E. Ideker, W. M. Smith, and V. G. Fast, "Optical measurements of intramural action potentials in isolated porcine hearts using optrodes," *Heart Rhythm*, vol. 4, pp. 1430–1436, 2007.

371. S. J. Shipsey, S. M. Bryant, and G. Hart, "Effects of hypertrophy on regional action potential characteristics in the rat left ventricle: a cellular basis for T-wave inversion?," *Circulation*, vol. 96, pp. 2061–2068, 1997.

372. T. Opthof, R. Coronel, and M. J. Janse, "Is there a significant transmural gradient in repolarization time in the intact heart?: Repolarization Gradients in the Intact Heart," *Circ. Arrhythm. Electrophysiol.*, vol. 2, pp. 89–96, 2009.

373. P. Taggart, P. M. Sutton, T. Opthof, R. Coronel, R. Trimlett, W. Pugsley, and P. Kallis, "Transmural repolarisation in the left ventricle in humans during normoxia and ischaemia," *Cardiovasc. Res.*, vol. 50, pp. 454–462, 2001.

374. C. Antzelevitch, W. Shimuzu, G. X. Yan, S. Sicouri, J. Weissenburger, V. V. Nesterenko, A. Burashnikov, J. M. D. Diego, J. E. Saffitz, and G. P. Thomas, "The M Cell: Its contribution to the ECG and to normal and abnormal electrical function of the heart," *J. Cardiovasc. Electr.*, vol. 10, pp. 1124–1152, 1999.

375. C. Antzelevitch, "Are M cells present in the ventricular myocardium of the pig? A question of maturity," *Cardiovasc. Res.*, vol. 36, pp. 127–128, 1997.

376. S. Poelzing, "Are electrophysiologically distinct M-cells a characteristic of the wedge preparation?," *Heart Rhythm*, vol. 6, pp. 1035–1037, 2009.

377. D. L. Weiss, G. Seemann, D. U. J. Keller, D. Farina, F. B. Sachse, and O. Doessel, "Modeling of heterogeneous electrophysiology in the human heart with respect to ECG genesis," in *Proc. CinC*, vol. 34, pp. 49–52, 2007.

378. X. Chen, V. Piacentino, S. Furukawa, B. Goldman, K. B. Margulies, and H. S. R., "L-type Ca2+ channel density and regulation are altered in failing human ventricular myocytes and recover after support with mechanical assist devices," *Circ. Res.*, pp. 517–524, 2002.

379. G. Malfatto, G. Beria, S. Sala, O. Bonazzi, and P. J. Schwartz, "Quantitative analysis of T wave abnormalities and their prognostic implications in the idiopathic long QT syndrome," *J. Am. Coll. Cardiol.*, vol. 23, pp. 296–301, 1994.

380. D. M. Roden, R. Lazzara, M. Rosen, P. J. Schwartz, J. Towbin, and G. M. Vincent, "Multiple mechanisms in the long-QT syndrome. Current knowledge, gaps, and future directions.," *Circulation*, vol. 94, pp. 1996–2012, 1996.

381. W. J. Karlon, J. L. Lehr, and S. R. Eisenberg, "Finite element models of thoracic conductive anatomy: sensitivity to changes in inhomogeneity and anisotropy," *IEEE Trans. Biomed. Eng.*, vol. 41, pp. 1010–1017, 1994.

382. R. McFee and S. Rush, "Qualitative effects of thoracic resistivity variations on the interpretation of electrocardiograms: the low resistance surface layer," *Am. Heart J.*, vol. 76, pp. 48–61, 1968.

383. P. C. Stanley, T. C. Pilkington, and M. N. Morrow, "The Effects of Thoracic Inhomogeneities on the Relationship Between Epicardial and Torso Potentials," *IEEE Trans. Biomed. Eng.*, vol. Bme-33, pp. 273–284, 1986.

384. Y. Rudy and R. Plonsey, "A comparison of volume conductor and source geometry effects on body surface and epicardial potentials," *Circ. Res.*, vol. 46, pp. 283–291, 1980.

385. Y. Rudy and R. Plonsey, "The eccentric spheres model as the basis for a study of the role of geometry and inhomogeneities in electrocardiography," *IEEE Trans. Biomed. Eng.*, vol. 26, pp. 392–399, 1979.

386. C. Ramanathan and Y. Rudy, "Electrocardiographic imaging: I. Effect of torso inhomogeneities on body surface electrocardiographic potentials," *J. Cardiovasc. Electrophysiol.*, vol. 12, pp. 229–240, 2001.

387. K. H. Haugaa, T. Edvardsen, T. P. Leren, J. M. Gran, O. A. Smiseth, and J. P. Amlie, "Left ventricular mechanical dispersion by tissue Doppler imaging: a novel approach for identifying high-risk individuals with long QT syndrome," *Eur. Heart. J.*, vol. 30, pp. 330–337, 2009.

388. B. Appleton, Q. Wei, S. Crozier, F. Liu, S. Wilson, L. Xia, and N. Liu, "An electrical heart model incorporating real geometry and motion," in *Conf. Proc. IEEE Eng. Med. Biol. Soc.*, vol. 1, pp. 345–348, 2005.

389. Q. Wei, S. Crozier, B. Appleton, L. Xia, F. Liu, S. Wilson, S. W., S. R., and R. R., "An MRI-based Beating Heart Model," in *Proc. Intl. Soc. Mag. Reson. Med.*, vol. 14, p. 1645, 2006.

390. M. Jiang, L. Xia, G. Shou, Q. Wei, F. Liu, and S. Crozier, "Effect of cardiac motion on solution of the electrocardiography inverse problem," *IEEE Trans. Biomed. Eng.*, vol. 56, pp. 923–931, 2009.

391. T. Arts, F. W. Prinzen, T. Delhaas, J. R. Milles, A. C. Rossi, and P. Clarysse, "Mapping displacement and deformation of the heart with local sine-wave modeling," *IEEE Trans. Med. Imag.*, vol. 29, pp. 1114–1123, 2010.

392. P. Kohl and F. Sachs, "Mechanoelectrical feedback in cardiac cells," *Phil. Trans. R. Soc. Lond.*, vol. 359, pp. 1173–1185, 2001.

393. T. G. McNary, K. Sohn, B. Taccardi, and F. B. Sachse, "Experimental and computational studies of strain-conduction velocity relationships in cardiac tissue," *Prog. Biophys. Mol. Biol.*, vol. 97, pp. 383–400, 2008.

394. P. Colli Franzone, L. F. Pavarino, and B. Taccardi, "Simulating patterns of excitation, repolarization and action potential duration with cardiac bidomain and monodomain models," *Math. Biosci.*, vol. 197, pp. 35–66, 2005.

395. B. J. Roth, "Electrical Conductivity Values Used with the Bidomain Model of Cardiac Tissue," *IEEE Trans. Biomed. Eng.*, vol. 44, pp. 326–327, 1997.

396. R. Modre, M. Seger, G. Fischer, C. Hintermüller, D. Hayn, B. Pfeifer, F. Hanser, G. Schreier, and B. Tilg, "Cardiac Anisotropy: Is it Negligible Regarding Noninvasive Activation Time Imaging?," *IEEE Trans. Biomed. Eng.*, vol. 53, pp. 569–580, 2006.

397. S. J. Walker and D. Kilpatrick, "Forward and inverse electrocardiographic calculations using resistor network models of the human torso," *Circ. Res.*, vol. 61, pp. 504–513, 1987.

398. A. van Oosterom and G. J. Huiskamp, "The effect of torso inhomogeneities on body surface potentials quantified using "tailored" geometry," *J. Electrocardiol.*, vol. 22, pp. 53–72, 1989.

399. S. E. Geneser, R. M. Kirby, and R. S. MacLeod, "Application of Stochastic Finite Element Methods to Study the Sensitivity of ECG Forward Modeling to Organ Conductivity," *IEEE Trans. Biomed. Eng.*, vol. 55, pp. 31–40, 2008.

List of Publications and Supervised Theses

Journal Articles

1. **D. U. J. Keller**, F. M. Weber, G. Seemann, and O. Dössel, "Ranking the Influence of Tissue Conductivities on Forward-Calculated ECGs," *IEEE Trans. Biomed. Eng.*, vol. 57, pp. 1568–1576, 2010.

2. **D. U. J. Keller**, O. Jarrousse, T. Fritz, S. Ley, O. Dössel, and G. Seemann, "Impact of Physiological Ventricular Deformation on the Morphology of the T-Wave: A Hybrid Static-Dynamic Approach," *IEEE Trans. Biomed. Eng.*, vol. 58, pp. 2109–2119, 2011

3. **D. U. J. Keller**, D. L. Weiss, O. Dössel, and G. Seemann, "Influence of I_{Ks} Heterogeneities on the Genesis of the T-Wave: A Computational Evaluation," *IEEE Trans. Biomed. Eng.*, (in print)

4. F. M. Weber, **D. U. J. Keller**, S. Bauer, G. Seemann, C. Lorenz, and O. Dössel, "Predicting Tissue Conductivity Influences on Body Surface Potentials - An Efficient Approach Based on Principal Component Analysis," *IEEE Trans. Biomed. Eng.*, vol. 58, pp. 265–273, 2011.

5. G. Seemann, **D. U. J. Keller**, M. Krueger, F. M. Weber, M. Wilhelms, and O. Dössel, "Electrophysiological Modeling for Cardiology: Methods and Potential Applications," *Information Technology*, vol. 52, pp. 242–249, 2010.

6. D. L. Weiss, **D. U. J. Keller**, G. Seemann, and O. Dössel, "The Influence of Fibre Orientation, Extracted from Different Segments of the Human Left Ventricle, on the activation and repolarization sequence: a simulation study," *Europace*, vol. 9, p. 96, 2007.

Conference Contributions

1. **D. U. J. Keller**, A. Bohn, O. Dössel, and G. Seemann, "In-silico Evaluation of Beta-Adrenergic Effects on the Long-QT Syndrome," in *Proc. CinC*, vol. 37, pp. 825–828, 2010.

2. **D. U. J. Keller**, O. Dössel, and G. Seemann, "Simulating Cardiac Excitation in a High Resolution Biventricular Model," in *Proc. BMT*, 2010.

3. **D. U. J. Keller**, R. Kalayciyan, O. Dössel, and G. Seemann, "Fast Creation of Endocardial Stimulation Profiles for the Realistic Simulation of Body Surface ECGs," in *Proc. IFMBE*, vol. 25/4, pp. 145–148, 2009.

4. **D. U. J. Keller**, O. Dössel, and G. Seemann, "Evaluation of Rule-Based Approaches for the Incorporation of Skeletal Muscle Fiber Orientation in Patient-Specific Anatomies," in *Proc. CinC*, vol. 36, pp. 181–184, 2009.

5. **D. U. J. Keller**, G. Seemann, D. L. Weiss, and O. Dössel, "Detailed Anatomical Modeling of Human Ventricles Based on Diffusion Tensor MRI," in *Proc. BMT*, vol. 50/1, 2006.

6. **D. U. J. Keller**, D. L. Weiss, O. Dössel, and G. Seemann, "Evaluating Mathematical Cell Models as to their Suitability for Modeling the Congenital Long-QT 2 Syndrome," in *Proc. BMT*, vol. 52, 2007.

7. **D. U. J. Keller**, G. Seemann, D. L. Weiss, D. Farina, J. Zehelein, and O. Dössel, "Computer based Modeling of the Congenital Long-QT syndrome in the Visible Man Torso: From Genes to ECG," in *Conf. Proc. IEEE Eng. Med. Biol. Soc.*, pp. 1410–1413, 2007.

8. **D. U. J. Keller**, D. L. Weiss, O. Dössel, and G. Seemann, "Transferring Ventricular Myocyte Orientation to Individual Patient Data Based on Diffusion Tensor MRI Measurements," in *Proc. CURAC*, pp. 255–258, 2007.

9. C. Otto, **D. U. J. Keller**, G. Seemann, and O. Dössel, "Integrating Beta-Adrenergic Signaling into a Computational Model of Human Cardiac Electrophysiology," in *Proc. IFMBE*, vol. 25/4, pp. 1033–1036, 2009.

10. R. Kalayciyan, **D. U. J. Keller**, G. Seemann, and O. Dössel, "Creation of a Realistic Endocardial Stimulation Profile for the Visible Man Dataset," in *Proc. IFMBE*, vol. 25/4, pp. 934–937, 2009.

11. S. Bauer, **D. U. J. Keller**, F. M. Weber, P. Tri Dung, G. Seemann, and O. Dössel, "How do Tissue Conductivities Impact on Forward-Calculated ECGs? An Efficient Prediction Based on Principal Component Analysis," in *Proc. IFMBE*, vol. 25/4, pp. 641–644, 2009.

12. D. L. Weiss, G. Seemann, **D. U. J. Keller**, D. Farina, F. Sachse, and O. Doessel, "Modeling of Heterogeneous Electrophysiology in the Human Heart with respect to ECG Genesis," in *Proc. CinC*, vol. 34, pp. 49–52, 2007.

13. G. Seemann, **D. U. J. Keller**, D. L. Weiss, and O. Dössel, "Modeling Human Ventricular Geometry and Fiber Orientation Based on Diffusion Tensor MRI," in *Proc. CinC*, vol. 33, pp. 801–804, 2006.

14. F. M. Weber, S. Lurz, **D. U. J. Keller**, D. L. Weiss, G. Seemann, C. Lorenz, and O. Doessel, "Adaptation of a Minimal Four-State Cell Model for Reproducing Atrial Excitation Properties," in *Proc. CinC*, pp. 61–64, 2008.

15. G. Seemann, S. Lurz, **D. U. J. Keller**, D. L. Weiss, E. Scholz, and O. Dössel, "Adaption of Mathematical Ion Channel Models to Measured Data Using the Particle Swarm Optimization," in *Proc. IFMBE*, vol. 22, pp. 2507–2510, 2008.

16. G. Seemann, C. Rombach, **D. U. J. Keller**, and O. Dössel, "Effect of the Drug Cisapride on Human Ventricular Cardiomyocytes: A Simulation Study," in *Proc. IFMBE*, vol. 25/4, 2009.

17. T. Fritz, O. Jarrousse, **D. U. J. Keller**, G. Seemann, and O. Dössel, "In Silico Analysis of the Impact of Transmural Myocardial Infarction on Cardiac Mechanical Dynamics for the 17 AHA Segments," in *Proc. FIMH*, 2011.

18. M. W. Krueger, V. Schmidt, C. Tobón, F. M. Weber, C. Lorenz, **D. U. J. Keller**, H. Barschdorf, M. Burdumy, P. Neher, G. Plank, K. Rhode, G. Seemann, D. Sanchez-Quintana, J. Saiz, R. Razavi, and O. Dössel, "Modeling Atrial Fiber Orientation in Patient-Specific Geometries: A Semi-Automatic Rule-Based Approach," in *Proc. FIMH*, 2011.

19. M. W. Krueger, K. Rhode, F. M. Weber, **D. U. J. Keller**, D. Caulfield, G. Seemann, B. R. Knowles, R. Razavi, and O. Dössel, "Patient-Specific Volumetric Atrial Models with Electrophysiological Components: A Comparison of Simulations and Measurements," in *Proc. BMT*, 2010.

20. M. W. Krueger, F. M. Weber, O. Jarrousse, **D. U. J. Keller**, G. Seemann, and O. Dössel, "Anatomical, Electrophysiological and Mechanical Modeling of the Heart," in *Chaste Users' & Developers' Workshop*, 2009.

21. M. Reumann, B. G. Fitch, A. Rayshubskiy, **D. U. J. Keller**, G. Seemenn, O. Dössel, and J. J. Rice, "Strong Scaling and Speedup to 16,384 Processors in Cardiac Electromechanical Simulations," in *Conf. Proc. IEEE Eng. Med. Biol. Soc.*, pp. 2795–2798, 2009.

22. M. Reumann, B. Fitch, R. A., **D. U. J. Keller**, G. Seemann, O. Dössel, M. Pitman, and J. J. Rice, "Orthogonal Recursive Bisection Data Decomposition for High Performance Computing in Cardiac Model Simulations: Dependence on Anatomical Geometry," in *Conf. Proc. IEEE Eng. Med. Biol. Soc.*, pp. 2799–2802, 2009.

23. M. Reumann, C. Morris, **D. U. J. Keller**, G. Seemann, O. Dössel, and G. R. Abram, "Towards Run Time Visualization in Cardiac Modeling," in *Proc. IFMBE*, vol. 25/4, pp. 999–1002, 2009.

24. M. Reumann, B. Fitch, A. Rayshubskiy, **D. U. J. Keller**, D. Weiss, G. Seemann, O. Dössel, M. Pitman, and J. J. Rice, "Large Scale Cardiac Modeling on the Blue Gene Supercomputer," in *Conf. Proc. IEEE Eng. Med. Biol. Soc.*, vol. 2008, pp. 577–580, 2008.

25. R. Miri, M. Reumann, **D. U. J. Keller**, D. Farina, and O. Dössel, "Comparison of the Electrophysiologically Based Optimization Methods with Different Pacing Parameters in Patient Undergoing Resynchronization Treatment," in *Conf. Proc. IEEE Eng. Med. Biol. Soc.*, pp. 1741–1744, 2008.

26. R. Miri, M. Reumann, **D. U. J. Keller**, D. Farina, and O. Dössel, "Computer Based Optimization of Biventricular Pacing According to the Left Ventricular 17 Myocardial Segments," in *Conf. Proc. IEEE Eng. Med. Biol. Soc.*, pp. 1418–1421, 2007.

27. R. Miri, M. Reumann, **D. U. J. Keller**, D. Farina, C. Wolpert, and O. Doessel, "Optimization of Cardiac Resynchronization Therapy Based on a Computer Heart Model Assuming 17 left Ventricular Segments," in *Proc. BMT*, vol. 52, 2007.

28. R. Miri, M. Reumann, **D. U. J. Keller**, D. Farina, and O. Doessel, "A Noninvasive Computer Based Optimization Strategy of Biventricular Pacing," in *Proc. CURAC*, pp. 133–136, 2007.

29. O. Doessel, G. Seemann, D. Farina, **D. U. J. Keller**, R. Miri, F. M. Weber, and D. L. Weiss, "Visions in Modeling of Cardiac Arrhythmogenic Diseases and their Therapies," in *Proc. IFMBE*, vol. 20, pp. 450–453, 2008.

Other Contributions

1. **D. U. J. Keller**, "Detailed Anatomical and Electrophysiological Modeling of Human Ventricles Based on Diffusion Tensor MRI," Diploma Thesis, Institute of Biomedical Engineering, Universität Karlsruhe (TH) 2006.

2. **D. U. J. Keller**, "Investigating Options for a Low Cost and High Performance Doppler OCT," Student Research Project, Optical and Biomedical Engineering Laboratory, University of Western Australia and Institute of Biomedical Engineering, Universität Karlsruhe (TH), 2005.

3. O. Dössel, D. Farina, **D. U. J. Keller**, R. Miri, M. Reumann, G. Seemann, F. Weber, and D. Weiss, *Modellgestützte Therapie*, vol. 13, ch. Modellbasierte Fusion von Herz-Bildgebung und Elektrokardiographie, pp. 225–236. Dresden: Health Academy, 2008.

4. M. W. Krueger, V. Schmidt, **D. U. J. Keller**, T. Fritz, G. Seemann, and O. Dössel, "Comparison of Methods for Visualization of 3D Myocardial Fiber Structure in Printed Images," in *KIT PhD Symposium 2010*, p. 80, 2010.

Supervised Diploma Theses and Student Research Projects

1. Stefan Bauer, "Efficient Reconstruction of BSPMs and Optimization of Multiple Tissue Conductivities," Diploma Thesis, Institute of Biomedical Engineering, Karlsruhe Institute of Technology, 2009

2. Carola Otto, "Impact of Adrenergic Regulation on the Electrophysiological Properties of Human Ventricular Myocytes," Diploma Thesis, Institute of Biomedical Engineering, Karlsruhe Institute of Technology, 2009

3. Pham Tri Dung, "The Impact of Distinctive Organ Conductivities on the Solution of the Forward Problem of Electrocardiography," Master Thesis, Institute of Biomedical Engineering, Karlsruhe Institute of Technology, 2009

4. Stephan Lurz, "Multidimensional Adaption of Electrophysiological Cell Models to Experimentally Characterized Pathologies," Diploma Thesis, Institute of Biomedical Engineering, Universität Karlsruhe (TH), 2008

5. Andreas Bohn, "Einfluss Beta-Adrenerger Regulation auf die Elektrophysiologie und Erregungsausbreitung im Menschlichen Ventrikel," Student Research Project, Institute of Biomedical Engineering, Karlsruhe Institute of Technology, 2011

6. Raffi Kalayciyan, "Einfluss des Endokardialen Stimulationsprofils auf die Morphologie des Oberflächen-EKGs," Student Research Project, Institute of Biomedical Engineering, Karlsruhe Institute of Technology, 2009